MORRIS AUTOMATED INFORMATION NETWORK
W9-BJV-193
R06001 56629

REPAIRING ANTIQUE CLOCKS

A Guide for Amateurs

Repairing Antique Clocks

A GUIDE FOR AMATEURS

Eric Smith

ARCO PUBLISHING COMPANY, INC
New York

Published 1974 by Arco Publishing Company, Inc.
219 Park Avenue South, New York, N.Y. 10003

Library of Congress Catalog Card No. 74-75397

ISBN 0 668 03440 8

Printed in Great Britain

Contents

Illustrations

PLATES

IN TEXT

Illustrations

(These drawings are not to scale)

Preface

This is a book for those who find clocks irresistibly attractive, no matter the reason. They may have seen clocks whose age and appearance delighted them, but which they felt unable to buy because of a label saying 'requires attention', or in their youth they may have dissected the family heirloom on the kitchen table and put it together again with less than complete success. Having read this book, they will still make mistakes, of course, though possibly not quite so many or so grave. But I hope that they will find their fascination and their horological ambitions stimulated to further attempts, however humble their objectives may be.

The libraries are reasonably well stocked in the field of clocks as antiques and works of art. There are also a number of excellent close studies on techniques of repair, though few of them seem directed towards the impoverished dabbler. There are, however, very few books which include in their pages something on both the antiquarian and the practical sides of horology.

It is my belief that for many people these two aspects of clocks—practical mechanical interest and antiquarian interest—are very closely connected. Yet these interests can conflict. I have no doubt that certain suggestions here for bringing an old clock back to life might be regarded by some readers as abuse of an object of beauty and a sin against the historical spirit. I respect these attitudes very sincerely and because of this I ask the reader to proceed with taste and caution. But, nonetheless, there is room for compromise, and the person who has bought an old clock has, after reviewing its age and value in the light of his own circumstances, a perfect right to do what he likes with it. How strongly he feels that he has to preserve a heritage, how well it may be preserved by professional restoration, how much better it may be preserved by his holding the clock incomplete and stopped—all this is his own affair.

Therefore I hope that the brief consideration of both sides of the subject which this book offers will be of interest both to those whose inclination is to one or other direction and to those whose fascination is, like my own, about equally divided.

1
The horological ham

It is touch and go whether you will find a definition of 'ham' in a dictionary and, if you do find one, the odds are against its striking you as very apt. The fact is that hams are a rather indefinable breed. They may belong to a club or society of like-minded people, but they have no special privileges, no access to public investment, and they find, as hams, no special kindness in their bank managers' eyes. They are often impecunious and are largely unqualified. If they take examinations in their chosen subject it may not be simply in quest of a professional accolade but rather an attempt to improve themselves, to make them better hams; and there are some who hold that even such a step is a confession that the ham is yielding to convention.

Hams are usually lonely and isolated beings. Their pursuits are individualistic and it is themselves they seek to please, although one sure way of pleasing oneself is to receive an acknowledgment of having imparted pleasure to another. They are free of any livery company or guild. They are lucky to be free of their wives. For their real life's work, they do not clock in and out of a concrete block alive with expensive machinery in which they have minimal interest. On the contrary, the ham busies himself on the table or, with luck, at a work-bench in the garage or in a diminutive shack at the end of the garden. Save that he must at a specified time go out and pursue his secondary interest as a bread-winner, the ham is not bothered by time and motion studies or like measures. It may take him two evenings to complete inadequately what the craftsman with a lifetime's experience will do very much better in half an hour. But no matter—so long as the job is done and, above all, he did it himself.

'Ham' is theatrical slang, and indeed, the ham is an actor. He did not exist in recognisable form until the old skills of hand and eye were systematised and mechanised for the mass-production of useful objects. In response to such a process there comes a feeling that the individual is increasingly denied the simple pleasures, physical and mental, of making, and the rise of the ham may be seen as the outcome of such a response. He is a humble imitator of the designer, the technician and the manufacturer rolled into one. But essentially, he serves no one but himself.

In his particular interest the ham is always exploring paths generally considered impassable or unprofitable, urged on by a desire to complete his work and to see the finished product. Time is very limited and in his endeavours he will use every available artefact. Then, when all the appropriations and improvisations have been made, what self-deceptions must take place, what trimmings be added, what minor adjustments be made, all to defer the moment when it is finished and there is nothing more to do. Ostensibly he pursues a normal and respectable existence, working by day, watching television by night, doing the garden or decorating at the weekend; but, inside, the ham knows it is a delusion of calm and that before very long he will have another 'idea' which he is compelled to pursue. He is not clumsy, or selfish, or a frustrated professional, but dedicated to making real what is within him. He does not really know why he is so dedicated, but he will devote the maximum possible time and energy to his particular interests.

The horological ham is one to whom clocks are a necessity.

When he sees a darkened shop window, preferably with cobwebs and not much paint and, in the gloom he descries a long shape and a round patch on top doing its best to catch the light, he crosses the road. In this he is much like the pure antique collector, but thereafter their ways part. If it transpires that here is a shining brass face showing something near the correct time and the case has a nice patina of polish on it, or worse, there is a label saying 'Gould' or 'Knibb' on it alongside some letters or other hieroglyphs more than three in number—then our ham moves sadly away. For he prefers to see a case with cracks in it or the top missing and a dial which plainly has not told the time for many years. He likes latent beauty and proportion well hidden by a thick plaster of filth and discernible, of course, by none but himself. Above all, he likes an unambiguous price-tag with no more than two figures on it or, preferably, no price-tag at all but a clean mark-up. This suggests the throwout, beyond repair, even the treasure to be had for a song, the objective of those who count value in happy hours to be spent in personal enjoyment, not in potentially profitable time wasted on the irreparable.

Then the horological ham is a haggler, though a haggler of a special sort. He deals, after all, largely in what is patently worthless; only he can visualise future possibilities linked with the splendours of the past, and he alone can place real value on what he buys—or so he hopes. It would be untrue and unjust to say he is not interested in Tompions and Grahams. Of course he is interested in them. Were one in somewhat tatty order to come his way, he would be delighted. But he is a realistic fellow. He knows that these things are for those who have money to burn, that they are destined for the homes of the wealthy, or the museums. He also knows that there are middlemen whose role is to guide these treasures to such places, collecting their profit on the way. They know well before the event that a Tompion or Graham will come on the market.

Certainly the horological ham makes deals and he may have a good head for business; but his approach and aims are humble. He has only to convince the dealer that what has been in that corner for years is as worthless as it appears to be. He is not interested in the piece which the dealer has bought for a trifle and intends to do up and sell at a large profit. So, by and large, it is not the clockshop, let alone the antique clockshop, that he frequents, but the junkshop and premises of those who buy up the contents of whole houses and must dispose of them quickly. He surveys the papers, particularly the local papers, every week, because very occasionally the honest bargain still appears there. Then he moves fast, for dealers do not waste time.

Nonetheless, there is no enmity between hams and dealers. Certainly there is competition, because neither will ignore a bargain, and very often the same bargain is in prospect. There are those dealers the ham never visits; they add a few pounds when they think a customer wants something 'on the cheap'. But there are also those who know that the dealer and the ham occupy different worlds which only clash from time to time, and then often by mistake. Dealers as well as hams are fallible and they would rather have a pound or two in compensation than a total loss. An ugly unsaleable calendar clock may be no use to a dealer who has not the time or even the contacts to find it a decent case; but the ham will find the 'works' of compulsive interest and may even devote the time to make an appropriate case. Again, many dealers buy in bulk at auctions; the make-weights may be of more interest to the ham than the actual item which the dealer wants and will keep.

It is worth getting to know the right dealers. You can do them a good turn from time to time when you hear of an interesting buy well beyond your reach; they in return will often inform you of an oddment nearer your range before it comes on the open market. It is no use resenting the middleman's cut in such circumstances; think yourself lucky to have friends. The professional antique buyer or dealer can scour the country while you are in the office and so save you a great deal of work. True, he and his fraternity are gradually depriving you of a real bargain but they are also providing a more stable source of supply. Moreover, although auctions, especially country auctions, are an excellent source, there is usually no chance for the ham to visit them during the week.

I have said that the horological ham is not to be found snooping around clockshops. Of course, that is only a partial truth, for there are clockshops and clockshops. While he has his fascination for all that ticks and will cast the occasional glance, I do not think that the average ham finds much to interest him in multiple jewellers. He may, as a tyro, conceive that there will be some bits and pieces of interest, somewhere at the back where they make the tea, but he will very soon find that this is not generally so, and that the staff are only familiar with the techniques of selling battery-wound alarm clocks and so on. Often the repairs are sent away and, even if they are not, those handled on the premises tend to be only those for which replacement units and spare parts are available.

I remember once being presented by a friend with a late nineteenth-century pendulum clock. It was nothing special, but a pretty article in an oak case with glass sides. He had picked it up whilst on holiday and found on his return that it neither went nor struck. He approached the high-street jewellers who said that they farmed that sort of thing out and would have to get an estimate. The estimate came in due course. After a fortnight the clock was returned, with the money, but also with the diagnosis 'beyond repair'. Presumably it was not worth their contractor's time. Brief observation revealed that it would not go because the hands crossed each other and it would not strike because one of the pieces was bent. I did not clean the movement, but making it run and strike properly took no more than twenty minutes.

There are, however, really interesting clockshops up and down the country although they are decreasing in numbers. You will normally find them in backstreets. They are often one-man shows of some antiquity, and the one man is usually to be seen crouching over a bench doing something delicate with a pair of tweezers or in an alcove at the back of the shop which is littered with clocks of all sorts and conditions. You can be quite sure he is not in business to sell clocks or watches but to repair them. He will occasionally sell one or two, but they are mainly second-hand or unclaimed repair work. He really does know his job and he can be an extremely useful ally—though rather disarming in his firm opinion as to what can and cannot, or should and should not, be done. He is a devoted craftsman and if he appears to have a chip on his shoulder it is because so few people seem to appreciate craftsmanship and how much it costs. He will not spend as long as you do on a job because he knows what he is doing but, nonetheless, his work is very slow and time-consuming; for he lives, and has always lived, with perfection.

It pays to approach these people with care and modesty. If you put your nose in at the door they scent either work or trouble, and you have to show them in the gentlest possible way that very likely you do not want either. You may, of course, want work done. You are often bound to rely on such people to do what you simply cannot do without much expense and time.

If you have had an accident letting down a mainspring or made a mistaken purchase and, in consequence, have a gear-wheel minus half a dozen teeth, it will be to such a shop that you will most probably take the unsymmetrical remains. Here they perform such miracles as cutting wheels which are beyond you. If this is the position, you have to be exceedingly careful. The inside of a clock has to such a man a sort of sanctity about it. He will not take kindly to your blundering about with what he has spent a life-time becoming acquainted with. Therefore you are virtually obliged to discard your normal honesty and frame a modest lie—it fell out, or you are bringing it in as a good turn for a friend. He will not believe you, naturally, but if he is the right man he will recognise and respect your tact. Probably he will say that sort of thing is not worth his while. Then you will persevere as if it were a great matter of importance and eventually he will tell you that you should have brought the plates and pinion. By this he means that he cannot make a new wheel unless he knows how long to make its axle ('arbor' in clock language), how closely it has to mesh with the next pinion in the gear train or the precise shape and size of the teeth. He also means that he will, with a little persuasion, do the job. Therefore you do not say that he has only to make an exact copy of the wheel you have provided; rather, you blush silently in deference to his craft and produce the plates and pinion from your bag. Then he accepts you as a friend.

On the other hand, it may be that you do not want work done. The reasons for a horological ham to gravitate towards a proper clockshop are profound and obscure. Sometimes he simply cannot be without the company of clocks and those who know about them. The time of those who run these shops is limited, even if it is not quite so limited as they tend to make out, and an oblique approach is often advisable.

I used to visit such an establishment in the Midlands. When I first saw it I took it to be a rather down-at-heel barber's shop. There was a sign for hair lotion in the unlighted window but in one corner I happened to see the weight of a long-case clock. There was nothing else by way of display. The owner had a very long and unpronounceable Polish name dully painted on the fascia. I peered through the door. The view was limited by a phalanx of long-case clocks, and I had to go in. Entry was not easy round the grandfathers but eventually I stood by the makeshift counter. There was an angled lamp of small wattage, a bald head, and a screwed-up eye glaring at some minute particle from a watch. The other eye had a black patch over it and the whole effect was somewhat disconcerting. I studied it for some time. A mechanical bell had announced my arrival and I imagined that that was all that was necessary. This was not so and later I learned that when you came in you stated your business or, if you had none, you fabricated some to start a conversation.

This man was a superb craftsman. He knew it and he had to be handled with appropriate respect. I soon discovered, however, that we had some affinity in that he regarded himself as condemned to the misery of repairing watches for a penurious clientele, whereas his real interest was in old clocks. It was a strictly professional interest and so was distinct from mine. On the matter of history he was very vague and I do not know that he ever sold an antique clock; certainly he gave no evidence that he ever meant to do so. I quizzed him on the long-case clocks more than once. We would approach the subject of a sale indirectly by a general discussion of their merits, coming round via their antiquity, as I thought, to the question of price. But we never made the whole course. He would tell me with rapture that one was a fine piece of burr-walnut, but of course you could not see that at the moment and it needed a lot of work done to it; that another had a most unusual striking mechanism, and so on. But if I ventured any non-technical question, judiciously feigning ignorance, such as how old they were, he merely said sadly, 'Very old, very very old indeed', and this assessment he applied equally to a squat and ugly Victorian one with a garish enamel dial and to a very fine London-made thirty-hour clock with the turned pillars of the late seventeenth century. Yet, in a queer way, he seemed to like our conversations, unproductive as they were. I suppose he was, as he said, deeply 'fed up' with contemporary indifference to his craftsmanship and, in addition, simply lonely for I never saw anyone else in his shop. I certainly enjoyed every minute; the place was filled with clocks of all descriptions and I liked to watch, as any ham will, to see how various jobs are done by those who know.

I never bought any item which he regarded as a clock but I did make one exceptional purchase from him. It was on top of his boxes of bits at the end of the counter one day. Like any ham, I find the prospect of a pile of odd parts at a knock-down price irresistible, and certainly such purchases have saved money over the years. I introduced the subject in an offhand manner at the end of our conversation as I was turning to leave, and asked him how much he wanted for 'that stuff'. 'Stuff,' he said in his foreign manner, 'my friend, you are right. That is some stuff!' It was difficult to see exactly what he meant, but I decided that he was none too pleased with the article in question. I was right. He carried on with a peculiar incoherent invective about the 'stuff'. He seemed to associate it with some tragic circumstances in his people's history. I hesitate to quote his words, but finally he said it was 'German muck. No good. Never any good. All "quatsch".' This was a promising estimate. I pursued the opening and emerged with the collection of parts covered in rust and old grease but, as always with him, well wrapped in soft tissue. When I got home I was surprised to find that here was an almost complete movement. Moreover, it was of the kind which lives under a glass dome, so the absence of a case was no disaster. It turned out to be a very old '400-day' clock, one of the first, complete with striking mechanism. It took me three months to clean it up and provide a base, a dial, glass dome, hands, and a few other minor pieces, but after that it went well until I sold it in a financial emergency.

This was, I think, a true ham's deal. The piece was, in the condition it was in, undoubtedly rubbish. He had abandoned it, and the bin-men were to collect it next week. I gave him 40p (90c) for it and I could see from the glint in his single eye that he was mightily pleased with the transaction. It was a rarity, not intrinsically valuable, being of doubtful reliability in construction and design, and distinctly disagreeable in appearance, but it was a collectors' piece and I sold it to a dealer, who seemed equally pleased with the arrangement, for over fifty times as much. (I say this not in pride—in all probability the dealer asked double my price for it—but to remind horological hams that it is not absolutely without fail that we make a loss when we are forced to part with one of our delights, although for me it is sadly near to the general rule.)

Besides being a haggler in his unassuming way and a haunter of obscure shop-windows, the horological ham is lost if he is not, by nature, a hoarder and a snapper-up of unconsidered trifles. There is not much to be said of this. It is uncontrovertible, and certainly will become so to you within a very short time if you are new to this way of life. It is one of the subtle distinctions between the professional and the ham (though there are those who are a bit of both). The professional has his stores, of course, but they tend to be boxed and labelled, row after row. They put the ham's shoe-boxes and plastic sandwich-containers to shame. But the purposes and styles of these collections are quite different.

It should be said here that the 'mass-produced' clock of modern times is very different from the 'hand-made' clock of former years. 'Mass-produced' requires some qualification. Clocks have been mass-produced at least since the early eighteenth century—when many parts were made in batches by specialist firms and assembled by a craftsman whose name (or often merely that of a retailer) was engraved on the dial. The best makers were often in fact firms employing many hands, each responsible for certain components so that, even when little or no contracting took place, we cannot necessarily distinguish the individual responsible for the whole or major part of a particular clock.

Nor can we in any easy way distinguish between hand and

machine made goods. The lathe is a tool thousands of years old. It requires manual skill in operation, and the automatic screw-cutting lathe is a relatively recent invention. But a lathe —or throw or turns, the clockmaker's manually operated lathe —takes time to set up for a particular job and it stands to reason that, once set up, it would be used for at least a rudimentary form of production run, turning a number of the same parts in a row before being re-adjusted for different work. Certainly, the further back one goes the greater is the chance that a clock was made for an individual and substantially by an individual as a one-off job, but it is clear that the boundary between individual craftsmanship in the narrow sense and limited mass-production is an uncertain one.

Perhaps the readiest distinction is between a gearwheel originally filed tooth by tooth by hand with later gears cut from rolled brass by some dividing attachment on a lathe but usually finished by hand, and the latest examples stamped out by huge presses from sheet metal or cut together in batches. The difference is usually plain from the higher finish of the 'hand-made' wheel, its general tendency to have more pointed teeth, its yellower, thicker, and less brittle metal, and its smoother edges.

You will find similar differences between screws. The older screws tend to be more rounded, their slots less square at the bottom, their threads rarely coinciding exactly with modern gauges and much less clear cut. The high quality nineteenth-century screw is highly polished or blued, the modern one tending to be rough and dull. The oldest screws normally have square heads and a 'v'-shaped slot, filed rather than sawn or cut in the head. They are very hard to replace; the 'v' is not a practical shape for the purpose and the modern screw-driver jumps out. Centuries of this wear are hard to imitate.

It is, of course, the older materials that the horological ham hoards. He is not interested in forgery as such, but he is interested in giving the right impression and he does not give the right impression if he bolts an ancient timepiece together with screws bought in a plastic pack from the local hardware shop. I do not say he never uses them. If he has no others he must use them but he tries to make them somewhat aged in appearance and he makes sure that they are in the least obvious places. He hoards his oddments because he knows that in many cases they are rare or very scarce. Plenty were made, but of those which survive most play a vital part in holding together clocks like the one which has come his way and which he hopes to resurrect. If it is true of screws, it is truer still of wheels and other parts. Some he may be able to use as they are when the time comes. Others will be invaluable for slight adaptation or as patterns from which to make his own replacements or to have them made. Finally, at the most pessimistic, these spoils always have a scrap value. New metal is expensive, tends to be less easy to work, and is often hard to obtain in the sizes and sections required. A new part for an old clock will always look and be better if it is made from old metal.

According to his conscience and what is available, so the ham's use of his accumulated bits will vary. He will rarely be able to spend money on the equipment for making up a good wheel—if he has money of that order to spare he will prefer, as a rule, to spend it on actual clocks. He can, if he has the right contacts, have a wheel made for him. But he may not have the contact and may not be able or willing to pay the price asked. He may not want to wait and, above all, he may want to do the job himself. He knows the size of the wheel required and he knows how many teeth it needs. He may find the exact counterpart in his box of bits. More probably he will find a wheel of the wrong size but with the right number of teeth. Then he will make the major decision to move the holes to accept the replacement wheel and, possibly, those of existing wheels, so that his 'spare' can be properly fitted. Provided that its teeth are the right size and shape, or a new pinion is also fitted to engage with them, he will find that this can sometimes be done satisfactorily. He will not publicise the fact that he has done it because he knows that there are those, whom he respects, who would regard it as an act of deplorable vandalism. But he is a ham, dependent on his junk-box, and he has his own very personal ideals and objectives. Perhaps he will leave the original holes. Perhaps he will try to block them in invisibly. It cannot be done. But he will not mind; he will have done his best and what he considers to be right.

Thus the ham does the best he can to finish with a clock of reasonable appearance and going to time. What is reasonable is a matter for his conscience, and consciences vary. He may not be satisfied until he has done a full restoration in period style. Then there are the innumerable gradations down to actual barbarities which would make a craftsman wince. But the ham is a free man. It is his clock and he can make what he likes of it subject to avoiding, as far as possible, that awkward situation where one improvises and years later learns with sadness where the true materials can be obtained and how the job can be done properly.

But the ham is in difficulties, notwithstanding his accumulation of spare parts, if he has no material supplier. Material shops are increasingly geared to watches rather than clocks and to replacement parts and units rather than to one-off jobs and specially-made reproduction items. The position, however, is not hopeless. Sometimes suppliers buy old firms and their stocks are, for a while, enriched with cheap tools, broken movements, parts, and outmoded materials. Sometimes, too, they are in touch with the skilled manufacturers the ham cannot reach and they can, at a price, get a really difficult process like cutting a fusee especially carried out. Nowadays such a firm does not need to be near at hand. Many of them operate with catalogues and by post and will send most requirements by return and even on credit. The disadvantage is that the ham's major need, to root through heaps of materials and tools and buy what takes his fancy as well as what he urgently requires, is not satisfied. It is, however, surprising what can be found in an assortment of illustrated catalogues and on the whole, with a growing box of spares, a few materials- and tool-suppliers' addresses culled from a directory, the picture is not too depressing.

Mention has been made of possibly delicate relations with dealers, and this raises the whole question of one's attitude to investment. I do not suppose that if, exceptionally, the ham is offered at an acceptable price a clock by, say Quare, he will turn it down any more than will a discerning and informed investor. Moreover, if such a piece came my way, I should be hesitant about doing much to it myself. Frankly, the stakes are too large. While I should have some pleasure in the beauties of the object and in the contemplation of appreciating assets, I should also be both worried and bored. The worry for the security of a unique thing—which no insurance could replace and which is therefore rarely seen—would doubtless be typical of most people. The boredom, however, would be rather more characteristic of the ham. A thing of beauty, doubtless, is a joy for ever, but for the ham it is very much more of a joy if he has had a hand in its restoration. His interest in the perfect purchase (or in the work which must, commonsense demands, be passed to someone more competent) is relatively slight.

Therefore the person who is captivated by interest in clocks has to make a fundamental decision as to where his emphasis is to be. He can specialise, it is true, in children's clocks, lantern clocks, lancet clocks, skeleton clocks or what you will and thus limit his activities to part of what is on the market. Certainly he can develop an interest in and devote a lifetime to a particular branch of the subject. All branches are together too vast to cover. But he has to decide where his real interests lie in the light of his resources—whether in investing, collecting the great names, collecting for comprehensiveness or oddity, repairing all that comes as a business on the side, or whether it is merely that it is a special pleasure to hear a new tick from an escapement which has not moved in orderly fashion for many a year.

Personally, I collect of necessity rather than choice or discrimination. I cannot bear to throw out a clock on which I have done some useful work and in whose present appearance I have had a hand. Broadly speaking, the longer has been its silence, the better pleased I am to hear its new life. I like to think of its maker or makers, but I am not greatly bothered whether they were the named and illustrious or were humble village locksmiths. One of my favourite clocks is a thirty-hour movement, one of the earliest long-case vintage, of great crudity and immense character. True, I have no idea who made it or for whom. I have to wind it every night, every pinion is shamefully worn. I filed up one of its wheels laboriously tooth by tooth, and its case I made myself. Its value, even to the simple collector of clocks, must be small indeed. But its value to me is very great. Did I not make the decision that either every wheel must be replaced and the original, but for the dial, be almost gutted, or else it must grind on unaltered? Did I not spend all those hours on the missing wheel and the case? Is it not now proportioned as I wanted it to be and with some semblance of its original form? Naturally, if I should come across a case suitable to its small and unusual dial and of appropriate period, I would consider using it. But such cases rarely come without movements of some sort, so my problem would not be solved and I would still have to 'make do' for one. These, to my mind, are the satisfactions and dilemmas of the horological ham. The question of investment arises, but it is rather secondary to his real interests.

Consequently, this is not a book full of illustrations of the author's collection. We all know books of that type. They have their uses, providing information as to what is 'right' for an era, what is the impression to aim at and, of course, as a source of information and inspiration as to what is excellent in clocks. But all too often they are not a positive incentive to the ham or even to the collector. For collecting clocks these days is big business and is growing all the time. We know from experience that what is nominally going, or has just gone for a song, will in reality cost several hefty oratorios. We know, too, that what is shown for the record as 'formerly in the Bloggs collection' is now displayed in some museum overseas.

This is not a book full of such illustrations because I do not have such a collection in any formal sense. It is a question of keeping what one can avoid selling. There are always clocks in the house making surprising noises behind visitors' backs and there is always bedlam when they start hitting bells and gongs somewhere around midnight. Each of them has its beauty or its rarity; but their abiding interest for me lies in the memory of when the dog died from sipping the cleaning solution, when toy soldiers of yesterday were melted down to form a pendulum bob or when, as the case received its final polish, the movement came off its seven-foot perch and took a nasty chip from the cranium directly below.

Horological hams love old clocks and they love working on them. Their anathema is not the despised 'cottage-style' long case with its chains bumbling noisily over the sprockets, or even the attractive French movement built into a case which does not belong and never could have belonged to it. What really pains them is the bracket clock on someone's hall table ('Oh, that! It's not gone for years. But it's very old, you know. Worth a great deal, I believe'); the silent face above complicated marquetry in the corner, the 'grande sonnerie' carriage clock, a wedding present which is uncertain of its ability to strike because it has never been allowed to try, all give concern to horological hams. They have no grumble with clocks as collectors' pieces or clocks as investments so long as clocks are seen, heard and, preferably, made to be going concerns. It is for those who react to clocks in these distinctive ways that this book is written.

2

Tooling up

TOOLS

From the viewpoint of the craftsman and the professional the ham horologist is, virtually by definition, a bad workman. He is, after all, subject to a multitude of accidents. He is not undertaking the same repair for the thousandth time in his life. A large proportion of the tasks which he sets himself are experimental—he knows the objective, but he has to discover how to reach it. He may or may not discover the proper way of going about things; it is satisfactory to him if he reaches the objective by any means at all. The craftsman who blames his equipment may very well be a bad workman, because he has the knowledge, the skill and the practical experience to do better. Moreover, he probably has the correct tools in the first place. For the ham, things are very different. If he blames his tools, he is most likely right. But the following evening he will be back at the same job with the same tools and the project will go like a dream. He may have discovered a better way, or he may simply have more luck. Still he may not buy any better tools and still they may well be the root of his trouble. Tools are just not very high on the list of his priorities.

But, if he cannot afford much, and will spend rather less, the ham still must have some of the basic professional equipment. He need not buy the most expensive items but he will usually be unwise to buy the cheapest. There are, the catalogues will tell him, tools which will make light and perfect work of almost any job which he is likely to face. He can buy a gadget for winding up mainsprings, which will save him cut fingers, a broken wrist, or a distorted spring. He can buy extractors of all sorts for the apparently simple task of removing clock hands. He can—many will say he should—buy sets, small or gigantic, of punches for various purposes such as riveting and pinning. He can invest in a lathe of whatever size he chooses (really he will need two lathes) together with all the paraphernalia which will at least double the price and without which a lathe is little use to him. He can have a patent oiler, or several patent oilers, which will deposit the required minuscule drop of oil exactly in the position required. If he likes the idea he can have a pocket bellows for blowing dust out of movements he has cleaned, or he can have a full-sized job for smelting and casting. If he wants to be busy cleaning he can invest in machines for the purpose, or, if finish is his ambition, then an electro-plating outfit will serve him well. There is scarcely any limit to the labour-saving equipment which he can lay in if he has the money and the inclination.

The fact is that most hams have neither. Gadgets, certainly, attract them as if by some natural affinity. Every time a new catalogue comes through the door one may well surrender to at least one gadget. It will turn up a few days later, probably not quite as imposing as expected, and then spend the rest of its life in the tool box or be adapted for some foreign use. For the ham, the jobs for which these gadgets are intended do not crop up very often. Only a few hundred clocks will pass through his hands in a life-time and the particular job will, in all probability, occur on only a few of them. The ham's attention is inclined to be focused on what *must* be done to make the clock a going concern, not on what *should* be done if it is to satisfy the Trade Descriptions Act or to have an A1 overhaul with the worries of a guarantee attached to it. If the job is a critical one and the clock does not continue to function as hoped, the ham is still able to strip it down and to have another go. He likes to be, but he does not have to be, right first time.

In these circumstances gadgets are expensive even if the right one is ordered and it performs correctly. In listing tools which the ham needs at hand, I am therefore going to concentrate on what he cannot afford to be without. Naturally, his own list will vary according to his ambitions and how far he has come (because the ham is for ever tooling-up for something new) but, nonetheless, there are essentials and he must decide for himself what else he absolutely must have and what he would rather improvise.

Screwdrivers

An old screw may look crude enough and beyond further mutilation, but it can be correspondingly tight to undo if well rusted. A more modern screw is liable to be highly polished, and may be countersunk, at least on a movement of quality. A penknife and even a kitchen knife can be used, but they tend to be hard on both the screw and the hands. Consequently, you need a fair selection of screwdrivers. One with a quarter-inch blade is fine for the larger jobs and for operating on cases. A short 'stubby' implement can also be a blessing, for many movements, especially the more modern, seem to have been inserted into their cases by a magician or a man with underdeveloped hands. Electricians' small screwdrivers I find extremely useful; when first bought they have blunt and often inaccurately formed blades, but with a little work with grindstone or file they can be shaped and sharpened as required. The novice will find himself constantly surprised by the smallness of concealed screws in large movements, and a set of jewellers' screwdrivers is a necessity at the outset even if you propose to tackle nothing smaller than a grandfather clock. Such a set can be bought very cheaply, and there is no reason why it should not be, but ensure that the blade is firmly fixed in the handle; a blade which refuses to turn the screw despite all action on the handle is a source of great annoyance. These drivers have a loose 'button' top; to use them with precision you place your index finger on the top, which stays still and gives control, and rotate the handle between fingers and thumb. All screwdrivers get blunt and mauled on occasion. They are easily reshaped, but remember that the object is to get the longest taper consistent with strength at the point. A blunt screwdriver tends to rise out of the notch in the screw. Remember also that you should imitate the craftsman when you can do so without too much time and expense. Therefore, whenever possible, use a driver of the same width as the screw head. One too small will spoil itself and the screw while one too large will either not engage properly in the screw's slot or mutilate the surrounding surface.

Taps and dies

When starting out, taps and dies for threading holes and rods are not essential. But you soon discover that when you have no replacement screw you have either to make one or to re-thread the existing hole to take a screw from your stock. A large range of taps and dies comes expensive. It is best to purchase a small assortment of BA sizes and then to supplement it as required.

Pliers

Good pliers, preferably several pairs, are essential; and brass-faced pliers for soft work and round-nosed pliers for bending are also handy. It does not pay to buy cheap pliers. They are useless if their joint works loose in a few months and it is vital that their jaws are properly aligned. Hold them up, closed, to the light and inspect for any marked gap between the jaws. If there is a gap, especially in fine pliers, it must be towards the handles and not towards the tip; when fine pliers are gripped tightly their jaws bend slightly and if they have a gap towards the tip they will allow the object being held to spring out.

Nippers

Like pliers are nippers, top-cutting or side-cutting, simple or of cantilever type for power. You will probably start with straightforward top-cutters and buy others as you need them. It must be said that the wire-cutters between the jaws of an ordinary large pair of pliers are no use for horological work—they are much too blunt and much too soft. With nippers it is false economy to be stingy. They must align perfectly, must be sharp and stay sharp. To get these qualities of accuracy and hardness you have to pay more for them, but it is well worth while.

Tweezers

The same remarks apply to tweezers, if you buy them. They give the ham a boost in morale but in much clock work are less than essential. You use them mainly on finer work like carriage clocks and for directing pivots into their holes between the plates of the movement. Nonetheless, they can be invaluable for holding a screw or a highly polished part and they do tend to reduce the number of journeys round the carpet on hands and knees. When buying them (you may as well have a large, and a really small pair) be even more careful than with pliers. You have to watch for lateral alignment as well as how closely the points grip together and how they react to pressure. It is possible to bend tweezer points so that they line up, but my experience is that a pair which has started crooked and been bent straight will always retain a delinquent tendency. It is easier to make sure at the outset.

Files

You need files of all shapes and sizes—and of all ages too.

New files are used on non-ferrous metals including brass and soft materials. Steel is much more easily shaped with a partially worn file. Files become clogged with swarf long before they become really blunt. They are easily cleaned by soaking in paraffin or petrol and then a good rub with a stiff wire brush. More obstinate filings can be removed with the edge of a piece of soft scrap metal. Files in the hands of the ham can be unpredictable and dangerous. The main danger lies in the fact that they are hard and very brittle, and usually they have a pointed tang instead of a handle. It is advisable to buy separate handles or an adjustable grip. Suitably cut pieces of thick dowel or broomstick make reasonable handles for larger files; heat the tang red hot and force it into the handle by tapping the butt end hard on the bench or table. It is also advisable not to use a file for anything except cutting. Used as levers, they tend to snap alarmingly, and special care is needed in using files to enlarge holes. But, properly used, a round needle file may safely be used to 'move' a hole.

It is not an easy matter to file a flat surface true. Eventually you acquire a knack of adjusting the pressure on the tool according to which end of it is on the object being filed, and it helps to think of a file as a kind of saw, for it cuts only on the forward strokes. You can knock up a sort of mitre-block for yourself, such as you would use for sawing angles but without slots in it, and clamp the piece to be filed against it at the top. File all three edges together and the result will be fairly good, though you have to true up the wooden guide from time to time. Alternatively, box-section steel can be used.

Files come coarse, medium, and fine, and you need them all, round and flat and of various widths. In addition a triangular file can be very useful for removing burrs of metal in awkward corners. Make sure that some of your assortment have flat toothless edges so that you can file one surface without spoiling an adjacent one. A flat-sided file tool, file on one side and burnisher for polishing on the other, will serve you well in finishing work.

Drills

You need a large supply of drills covering a fair range of sizes and of various types because, in this work, drilling is rather an experimental process. Flat and spear-headed drills can be made but I doubt if the average ham makes a success of it or finds it worth the trouble even though the tool books recommend it. They can be bought quite cheaply in assortment, as can the slightly more expensive fluted and twist drills which are easier and quicker to work with, and tend to keep their edge longer. The wood-worker's hand-drill, let alone the brace, is not suitable for many jobs on clocks, except the occasional work on cases. It is essential to buy a drill stock, sometimes known as an archimedian drill, or else to use the power of a small motor coupled to a flexible drive and chuck. An ordinary hand power-drill is little use. It is too heavy and there is too much play in its bearings and chuck.

The ham will very soon discover for himself that, however urgent is his particular need, it pays to take time and trouble when drilling clockwork. Even at the lowest level of craftsmanship the mark of a drill which went astray is not a pretty sight. A drill broken in the work is exceedingly difficult to remove, since it is by nature as hard as the drill with which you try to drill it out, and, once extracted, it will leave a hole a good deal bigger than was wanted. The wise course is never to attempt drilling without marking the spot with a sharp centre punch, thus providing a foothold, and, for any but the most microscopic holes, first to drill a pilot hole or even a series of pilot holes of gradually increasing size until a drill of the correct size will give no trouble by binding in the work.

Broaches

The proper tool for enlarging most small holes is not a round file but a broach. You need an assortment of broaches, which you can buy by the pack. They are tapered, pentagonal lengths of hard steel with sharp edges. They usually end in sharp tangs like those of small files and are dangerous for the same reasons. For broaches, however, you *have* to have handles, or an adjustable pin vice, in order to make any use of them at all since they cut by being turned in the metal. Remember always that a broach is tapered; you must try to keep it upright, and if you use it from one side of the metal you must also use it from the other or your hole also will be tapered. For some jobs, notably bushing holes and making a hole for a pin, the taper is exploited and you broach mainly from one side.

Punches

Many a pin is not well-fitting, and many break in the process of extraction. Knocking broken pins out from the rear is one of the many uses to which punches are put. Others are marking for drilling, closing up existing holes, and riveting. You can make punches, though it is difficult to do so accurately without access to a lathe, and you can improvise. Doubtless many a professional, and certainly many a ham, has had recourse to a large nail in default of the appropriate punch. However, hard and sharp punches for marking are a cheap and useful aid. Hollow punches, used for riveting and for closing holes, can also be made, but again the truth and hardness of the commercial article is worth the money.

It is not easy to hold and hit a punch perfectly vertical and unfortunately it is very often necessary to do so. The ham, as usual, makes his decision and compromises where it suits him. You may make do with half a dozen assorted punches and trial and error, and perhaps if you do not attempt too much you will be lucky. On the other hand, you may find a constant succession of clocks passing into your house, especially of the smaller and more delicate varieties where tolerances are fine. Then perhaps you will dig into your pocket and buy a more or less expensive 'staking set', which will comprise a tool for holding punches dead vertical above one of a series of graduated holes, together with a large selection of punches. You will certainly marvel at what can be done with this tool and take all steps to keep rust at bay, but whether in the end you will consider it worthwhile depends on the scope of your operations. Whatever you decide, you will certainly want a well-balanced and light-weight hammer to hit the punches, and a steel stake with graduated holes in it is fairly essential. Ideally, the stake should be split down the middle between the holes and screwed together; it can then be placed round an arbor which has, say, a pinion on one end and a wheel on the other. This kind of stake seems, however, to be regarded as antediluvian and you will probably have to make your own rather than buy one.

Saws

It goes almost without saying that you need saws. You can do with a large hacksaw and a 'junior' model, and also a large supply of blades (old blades can be put to use as specialised files and, as with files, the sharpest blades are reserved for brass and soft materials). A saw of the fretsaw variety is also useful and so are a couple of pairs of tinsnips, one small with curved blades and one more massive.

Buffs and burnishers

Burnishers are extremely fine files, charged with an abrasive and a lubricant for polishing steel, and can be made of old files ground flat. But of greater use in the more humble workshop are emery sticks and buffs. You can buy these, but in this case I recommend making them. Use strips of wood about nine inches long and stick firmly and smoothly around about five or six inches of them a coating of emery paper or cloth. You need as many different grades as you can get, from coarse down to the finest. This lowly tool is invaluable for putting a reasonable finish on a previously rusted or now repaired part but should be kept away from the fine polished pivots of French clocks, which it will only damage. Equally valuable is a series of sticks—various shaped mouldings serve very well—covered with 'chammy' wash-leather. You can get a much better rub onto polished brass with one of these than you can with a sheet of wash-leather wrapped round your finger, and you can clean them easily with soft soap or washing-up liquid. Another improvisation worth considering is the sort of emery paper sold in cosmetics departments as hair removers—it is extremely fine and very cheap.

Brushes

A variety of brushes is also needed for cleaning. You can do with medium and fine clock-brushes, but it is cheaper to supplement them with old tooth-brushes, bottle and razor-cleaning brushes, and the like. Small steel and brass brushes of the type used for cleaning suede shoes are useful accessories.

Power unit

I have mentioned, in connection with drilling, the use of a small electric motor. It is, of course, in no way essential but it is extremely valuable where a lathe is not available and it is relatively cheap. You can buy an assortment of burrs and buffs for shaping and polishing metal and you will need a flexible drive and chuck to hold them. Such motors can be bought with a grindstone attached, which has obvious uses and, perhaps the most versatile and humble gadget of all, a brass scratch brush can be fitted to the other end. It will not produce a high finish, but it is unequalled as a quick remover of adhesive muck without damaging the part concerned. Rarely can a rotary steel wire brush be used—it is altogether too brutal. In the chuck of this power unit you will be able to insert drills. You have to be careful here, especially with steel, not to let automation go to your head; if you run the apparatus at too high a speed you will only succeed in burnishing the metal to a highly polished and impenetrable pit rather than in drilling your way through it. You can also fit a fine-toothed circular saw blade to the motor and make light weight of the smaller cutting jobs but only do so with extreme caution. Of all the items which are, as it were, the second stage of tooling up, rather than the bare essentials, I think such a power unit is perhaps the most important and will bring the quickest dividends.

Vices

A vice of some sort is essential—a two-inch engineers' vice will do well. Generally speaking, the thread on the screw of a woodworking vice is too coarse for the adjustments of tension which you need to make and often the jaws do not line up well enough. Of course you can go out and buy a variety of elaborate grips and vices which adjust to various angles, release at the flick of a lever, fix to the table with suction pads, and do other remarkable things. Do so if you wish, but remember that the first requirement of a vice is that it grips hard without doing damage; if you can buy the sort with a choice of hard or soft jaws, or to which protective blocks can be fitted, so much the better. By popular repute, the place for the clock man to work is 'at the bench', and certainly a bench of comfortable height soon becomes desirable. But for the start, a good solid table with, of course, a slab of wood over it to protect it if she insists, will do very well. Whichever it be, you will not want to be restricted to working on it all the time, even supposing that your lighting system is perfection itself. You can make good use of a small hand-vice, pintongs (a minute vice on a handle), and one or two pin-vices (compression chucks on handles) are helpful in holding the small round objects which have to be worked. Be careful in buying a hand-vice or pintongs that the spring is not too tight; you do not want to have to tighten it with a pair of pliers, and ideally you need to be able to feel how much pressure you are exerting on the article held.

Glasses and lighting

Finally, you must be able to see. If you insist on an eyeglass, by all means have one, for it can be good for the morale. But you will not find much use for it except with the finer French clocks, carriage clocks and their escapements. One with a focal length of two inches is about right. If you yield to this temptation, try to cultivate a relaxed sort of squint through it with both eyes open; it is very wearing to stare through an eyeglass with one eye and you will soon find the eye waters so that you have to take the glass off again. There is a variety of gadgets for attaching an eyeglass to a pair of spectacles and some of them have a very impressive appearance. Whether or not you have a glass, however, you must have good lighting. Daylight is best, of course, but as a ham, you will, perforce, do a good deal of your work at night. An adjustable lamp is then virtually essential. A fluorescent light is good for less local work, but there are some who find a perceptible flicker distracting and a source of headaches.

MATERIALS

When it comes to materials, it is difficult to specify an adequate list for someone who is setting up. The list can be very long, if he proposes to cover most contingencies in advance or it can be very short if he proposes to do the basic jobs and assemble the materials for more special work when the occasion arises. My inclination is to follow the latter principle; not all materials keep well, and I find there is an unavoidable tendency in my ménage for nice pieces of wire, sheet metal, or well-chosen strips of moulding for a case, to be used to make toy winches for children or to batten down roofing felt over a coal-bunker. Moreover, a large stock uses valuable space.

Most of the materials you need can be obtained from a good ironmonger or tool shop, so far as actual metal-work is concerned, and from the chemist so far as cleaning and finishing are concerned. Do acquaint your chemist with your purposes; chemists are sympathetic allies and they often have downstairs or out the back, small bottles of chemicals, pastes, and so on, of which you very likely have no knowledge at all, or which you believed to be extinct. But they are also, and rightly so, responsible people; they want to be assured that you know what you are doing with acids, silver nitrate, and other dangerous and corrosive substances. Your best bet with a chemist who is new to you is to buy a pair of rubber gloves from him as a starter; you must have and use rubber gloves, and their purchase will start you off on the right lines.

Oil

The first essential from the materials shop is oil. You need machine oil and grease for your tools and for lubricating in some drilling and cutting jobs, but you need clock oil (or watch oil) for clocks, and this is an instance where it is folly to try to make do. Tool oils go hard and sticky in a very short time when used in small quantities. They not only stop a clock by clogging its action but they collect dust and dirt which make a good abrasive and will ruin the polished acting surfaces. Vegetable clock oils vary, but almost any will do for your purpose and one small bottle will last you for a year or two provided that you keep it well sealed and do not dip dirty tools into it.

Pins

Clocks of any antiquity and dignity are commonly held

together by tapered pins driven into holes. You can buy these pins or make them. You can use old ones, though in general it is not recommended because they are difficult to clean and straighten and if you are going to this trouble you may as well make new ones. To make a pin you use brass or iron wire as required (usually iron pins in brass holes, and vice versa, but often iron pins right through), holding it at a slight angle on a wooden block with a groove in it a little smaller than the size the pin is to be; turning the wire continuously, you file until you have the required taper. As far as possible you avoid flat sides on the pin, but notice that one side of the pin which holds a hairspring in its stud is flattened so as not to distort the spring against which it presses. Personally, I am in favour of buying pins by the gross and only making the odd one when out of a particular size. You will of course come across many sorry adventures with tailors' pins and panel pins and sewing needles and you may be tempted; but you will find that none has a sufficiently gradual taper to hold firm and in addition needles are too hard, break easily and fail to provide the friction required for gripping. Pushing home a well-fitting pin which grips hard and yet comes out when required is one of the simple pleasures one should not be tempted to miss.

Keys

Not every clock you obtain has a key with it, and many clocks have the wrong keys. Keys of the wrong size wear out quickly and will spoil their winding squares. Buy a range of keys or, better, two sets of sizes mounted on a centrepiece (a 'star' of keys), then you will not be caught out; you can use it temporarily, note from it the size required, and then order the right one. In due course you will, naturally, accumulate an assortment of keys.

Pegwood

Pegwood is bought at the materials shop. It is used for cleaning out holes and recesses. Matchsticks will do initially or in an emergency, but true pegwood is cheap and very much better, being stronger and harder.

Shellac and solder

The clockman's adhesive is shellac. You can buy it in sticks and crush bits off them, or buy it in flake or crystals. It is an extremely strong, brittle adhesive, and a little goes a long way. The main use is for temporary fixing of small parts onto a larger surface to simplify working on them, but it has more permanent employment in sticking escapement pallets in position. Soft-solder is not much used by the craftsman, being ugly and not very strong. It also prevents the successful use of silver solder at a later stage. There are, however, many occasions when the ham makes use of it, particularly in concealed places. The rule is not to use soft solder when it can be avoided—and naturally in time one graduates away from it. Silver solder must be used in delicate fixing jobs where a very strong joint is required—one of the principal examples is in repairing clock hands. If you buy silver solder, buy the grade melting at the lowest temperature; a small blowlamp or a butane gun will produce a sufficient temperature for this. To make any solder flow, flux is required. Soft-solder is available incorporating flux, but borax or a borax-based flux has to be applied to the surfaces on which silver solder is to be used.

Enamel and lacquer

While you are in the materials shop, or buying from the catalogue, it will be as well to purchase a bottle of blue steel enamel. The rich blue finish of properly blued screw-heads and hands is a pleasure to look on against surrounding brass, but this blueing can be a chancy operation for the ham to carry out, and the blue enamel is worth having by you at least as a standby. Again, the books on how these things are, or were, done properly usually contain recipes for good clear lacquers, which are applied to visible polished brass parts (but not wheels or working surfaces) and cases to prevent tarnish. These days, you will find a bottle of clear lacquer at the materials shop which will do the trick more cheaply and as well. Lacquer can also be obtained in an aerosol and this can be applied so as to give a pleasant, slightly dappled finish to brass cases.

Metal

Your stocks of metal depend on how much manufacture of missing parts you are likely to undertake. I suspect that the general practice is to make few, if any, parts until a fair number of movements which are irreparable by virtue of wear or unhappy accident have passed through your hands. These you will naturally have kept and they will supply you with scrap metal which is, in most cases, of better quality and more suited to your purposes than new materials. However, a stock of brass, iron, and blue steel wire, in various thicknesses, is always useful and a sheet of brass of about 20 guage will not come amiss.

Bushes

Finally, from the materials shop, you are almost bound to need bushes for closing up worn pivot holes. You can buy them assorted and ready-cut or by the length, which you cut yourself. You need a large assortment; worn holes do not come in predictable sizes. The clock man works to fit rather than to standard measurements—and the smaller bushes have a habit of gravitating towards the floor.

Chemicals

From your chemist, or your general stores, you need a good metal cleaner; you will very soon decide with which of the many varieties you prefer to work. The 'long-term' type works in that it does keep tarnish at bay, but it is less effective on really filthy brass and is best borne in mind as an addition, a finisher, rather than as the main polishing agent.

You need ammonia to mix with soap for basic cleaning. Any proprietary clock cleaning fluid will do as well, but little better. Benzine or petrol is used to rinse off the ammonia. A powerful rust-remover is a necessity for the older clocks, especially those with iron frames, if they have been without a case for some time. It is worth experimenting with the various brands available which prevent the formation of new rust as well as removing the old. They are all dangerous in contact with the skin as, of course, is hydrochloric acid which can be used for the same purpose when all else fails. (I say as a last resort partly because it is so dangerous to have in the house and partly because articles stripped with it are not easily returned to a pleasing appearance.) Whether with rust removers or acids you should consult your supplier as to means of neutralising and removing such substances.

Turpentine (not turpentine substitute) is used as a lubricant and a coolant in drilling steel. It prevents binding of the bit in the metal and it will also help avoid (with the drill going at moderate speed) the burnishing of the steel so that it cannot be cut away. The other major use you will find for turpentine is as a solvent of beeswax (which you can also buy at the chemist's) in producing a nice deep wax polish on a wooden or marble case.

This completes the primary list of tools and materials for which you will find an urgent need. Some, as has been indicated, are more important than others. The ham sometimes yields to irrational enthusiasms and buys costly and rarely-used gadgets. He also tends to improvise on the basis of what is available. Only you can judge, in the light of experience, when the limit of improvisation has been reached and the proper tool must be purchased but it is worth bearing in mind that money spent on really good quality standard tools is rarely wasted in the long run. They not only perform better, but they also help the workman to perform better than he would otherwise have done.

3

What is a clock?

THE GEAR TRAIN

Most clocks are motors. They store energy put into them from an outside source and release it at a controlled rate. They go on releasing it until there is no longer enough left to move the inert mass of metal which their gear-wheels constitute, or until, though the mass will still move, it will not move reliably enough to be satisfactorily controlled.

You may, when working on a clock, think of it from either end. You may start where it is wound up and work through to the fastest moving wheel, or you may start with the fastest wheel and work back to the slowest one. One has to keep the basic concept clear. Remember that the source of the energy is you, and you apply it as you wind up the spring or raise the weights against the force of gravity. The other end of the clock—the pendulum or balance-wheel—is not a source of energy, but the controller of the rate at which the stored energy is released. Only in a very few clocks is the controlling agent also the source of the energy, the initiator of the proceedings. I am thinking principally of electric clocks where the pendulum may, by means of a pawl and ratchet, push a wheel round, or where the armature of an electric motor, its speed controlled by the power-station, drives the gear train. In these exceptional cases energy is applied to the controller which itself drives the clock. But for our purposes it is best to think of energy going in at one end and being released in a controlled manner at the other.

Thinking of a clock as a motor, it may occur to you that if there were a long enough train of gears that would be all that would be necessary. Provided the sizes of the gears are properly arranged, the faster the wheel at one end revolves, the slower the wheel at the other end will revolve. You know that it takes you perhaps six turns of your wrist to wind up a clock every twenty-four hours and that a clock's hour hand revolves twice in the same period; so clearly it must be arranged that some of the wheels go round more slowly than the wheel to which the spring is directly attached. You know also that the minute hand goes round twenty-four times and the second hand, if any, sixty times as fast again, so clearly it must be arranged that some of the wheels go round faster than that wheel with the spring on it. All this can be done by correctly calculated gears and, if the right strength of spring were involved and could be trusted to exert a constant force throughout the day, day after day, your clock could theoretically be a very long train of gears and no more; the last wheel in the train would go round very fast indeed, and the hands would be fixed to the axles or arbors of wheels going round at the right speeds somewhere in the middle.

This is, in practice, difficult to arrange. For one thing the energy dissipated in turning so many wheels is enormous and, for another, springs of such size are very much more powerful when fully wound than when half-unwound, though there would not be the same problem with a massive weight which would fall with constant force until it reached the floor. A further snag is the variations in friction which occur in gearing and which in a train of such length would cause great inaccuracy. Nevertheless, there have been modifications of such a system, or attempts to create the illusion of it. A very popular little novelty clock was made in the Victorian era which was just such a motor. At the end of the gear train, the arbor of the last wheel was extended into a wire crank. In front of the simple cardboard dial there was a projecting arm from which hung a steel ball, the size of a marble, on a piece of silk. This ball on the string was caught by the revolving crank and caused by it to revolve. Compared to the small last wheel, the steel ball was relatively massive; it slowed the rate at which the wheel revolved. This rate could also within limits be controlled by altering the length of the silk thread and, thereby, how large a circle the marble made. This was an attractive curiosity under its little glass dome. The works were fully visible, but the spring was concealed in the wooden base, so how the motor kept going was a minor mystery. How it kept time was a larger mystery. In fact it kept exceedingly bad time and could not but do so. For, while the ball on its thread acted as a somewhat imprecise centrifugal governor, which speed forced to rise and therefore to slow down, the whole apparatus was inordinately subject to the characteristics of the spring which powered it. In consequence it ran fast for half the day and slow for the other half, and you adjusted things until the time was right when you wanted it to be.

This was a motor and it told the time after a fashion but whether, for our purposes, it was really a clock is difficult to decide, just as it is difficult to decide whether the horological ham will include synchronous electric clocks (which are electric motors controlled by the frequency of the mains supply) among his interests. I leave aside the fact that a 'clock', strictly speaking, strikes the time on a bell (a 'cloche' in French, a 'Glocke' in German), and anything which merely points silently to figures is a 'timepiece' rather than a clock. Personally, I think we should define a clock which would be of interest to us as a motor where the final speed is governed by an agent which performs constantly, irrespective (as far as possible) of the energy applied to it and of the distance which the energy, slowly released, obliges it to travel. Such an agent is not only reliable, but it reduces the problems of that enormous train of gears continuously revolving, because it interrupts the motor, stopping it at repeated and regular intervals. And, again, to satisfy us a clock must store the power which from time to time is given to it; those electric clocks (not all electric clocks) where the power is continuously applied from outside do not appeal in the same way at all.

THE ESCAPEMENT

The interrupting device of a clock is known as the escapement. It is in essence no more than a small metal arm which drops into the progress of a toothed wheel and stops it and is then removed to release the wheel again. The regularity with which it does so is controlled by a pendulum, or a hairspring and an oscillating wheel, and these, since they not only swing, but also swing back again, arrange for the little interrupting arm to jump out of the toothed wheel and allow the motor to move again until the arm once more drops into its way. You can see why you must, at the outset, have a clear notion of where the energy goes in and what parts do the work in a clock. The escapement itself produces no energy and of itself drives nothing. Although it is where the weakest power is, it is associated with the fastest wheel of the clock's train. It is an interrupting device to a motor which, without it, would run down in a relatively short time according to the strength of its spring and the length of the train of wheels. It works on the last wheel so that the interruption can occur frequently (otherwise an enormously long pendulum would be needed) and without great violence.

It is the property of a pendulum, which is theoretically a mass at the end of a weightless thread, to take almost the same time to swing through a short arc as it does to swing through a long arc, provided always that the length of the pendulum remains the same. The hairspring attached to a balance-wheel has the same property—it is a substitute, in eventually

Page 49 A typical Vienna regulator

Page 50 Thirty-hour clock (c 1685) in home-made case

returning the wheel to the same static position, for the constant force of gravity which will eventually bring the pendulum to a stop. I have mentioned how the power of a spring is far from constant as it unwinds. A falling weight, such as drives a long-case clock, is very much more constant but, however perfectly the clock is made, there will still be variations in the power transmitted through its movement, due to minute imperfections in machining, to the compromises which have to be made in design to secure strength and due, even, to atmospheric pressure and temperature variations. The pendulum and the balance-wheel and hairspring, possessing this property of 'isochronism', are the means of surmounting the difficulty of inconstant forces which, uncontrolled, would ruin all attempts at time measurement.

But now, you will see, there is a further problem. Let us suppose that there is an arm attached to the pendulum (as there was in the earliest designs for the use of a pendulum), and that this arm, as the pendulum swings to and fro, repeatedly interrupts the turning of a wheel. In due course the pendulum will come to a stop, and either the motor will come to a stop as well, or it will run on until all the power is gone, depending on where the arm is when the pendulum stops (this position remains important even in the perfected escapements, as we shall see later when considering setting a clock 'in beat'). The same will happen with the balance-wheel. Therefore the matter has to be so arranged that the motor gives the pendulum or balance-wheel a little push (known as the 'impulse') to keep it swinging. This is most conveniently done at every swing, and it is achieved by varying the shape of the interrupting arm (or 'pallet'). If a straight face comes into the path of the wheel, it will stop it, but if the edge of the pallet is angled so that as it swings up and out of the wheel the tooth of the wheel slides down it, the effect is that the tooth tries to push the pallet away and this push is the impulse which is given to the pendulum. Alternatively, or in addition, the same effect can be given by using a straight pallet face and an angled tooth. Though the pendulum or balance is relatively substantial compared with the escapement wheel (or 'scapewheel') which keeps it going, the required impulse given can be very small indeed, because it occurs without fail at every swing.

In principle, then, the clock is essentially a motor driving a series of step-up gears and further controlled by accurately spaced interruptions in the form of an escapement. As one tooth of the last wheel in the train, the scapewheel, is allowed to 'escape', it slides down the straight face of the pallet until it reaches the angled part which gives the impulse, and then the next tooth falls onto the straight surface and is locked there. This happens every second in a long-case clock and of course much more frequently in smaller clocks and watches. The principle is the same, but of course the details vary from period to period—there may be a pendulum or some sort of balance-wheel, the pallets may be angled steel surfaces or steel pins, or they may be shaped jewel stones in brass mounts, a shaped tooth may fall on the back of half a cylinder and be stopped, and then, as the open half swings round (being part of the balance-wheel), it will push itself against the edge, give the balance a push, and then pass into the cylinder and be locked there until the wheel turns and lets the tooth out. There are still other forms of escapement, and the collection of them is one of the enthusiast's joys, just as it is one of the ham's delights to work on one he has not come across before. The principle of a sudden interruption and then of a slide along an angled face to give the impulse is universal and if you recognise it you will be able to understand most escapements which you come across (see Fig 1).

From this outline, you can see that the motor portion of the clock has really no purpose but to keep itself going and to keep its controller in action. So long as it does this, all is well; provided things are adjusted so that the escapement can perform as nearly as possible regardless of the force applied to it,

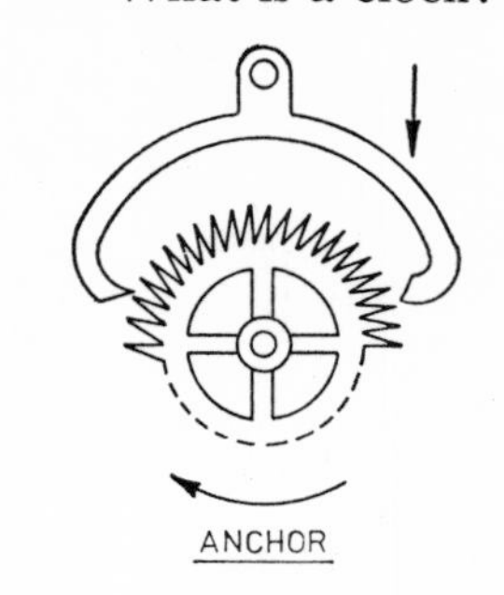

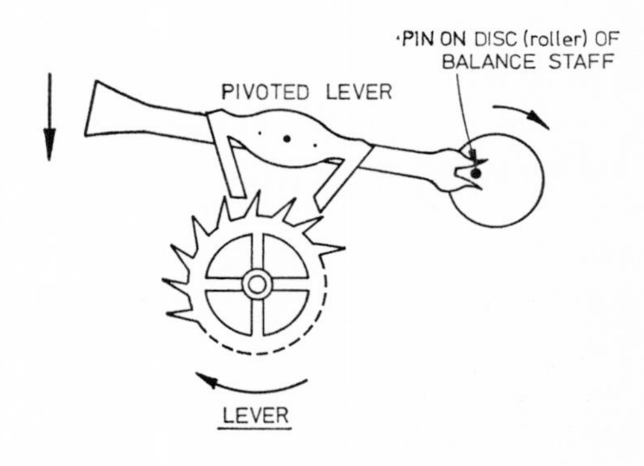

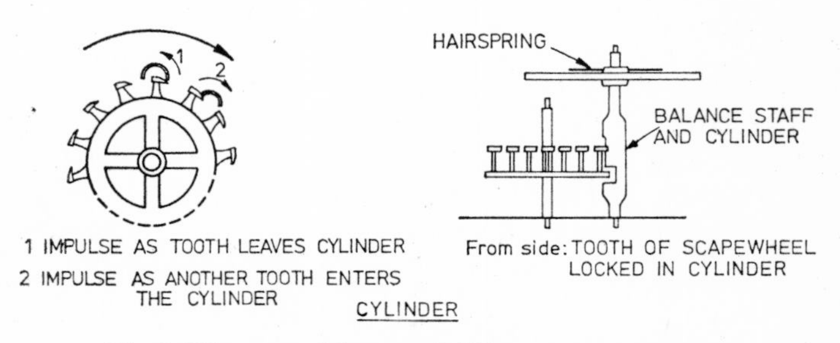

Fig 1 The resemblance of different escapements

the clock cannot fail to keep time so long as the pendulum or hairspring is the correct length. This perhaps is where the artist and the mechanic unite in the clock ham. The motor which has no purpose but to keep itself going has a certain aesthetic appeal, superior to the motor which washes the clothes. The smallness of the force which is regularly exerted on a mass the size of a long-case pendulum to keep it in motion is another source of mysterious pleasure. I never begrudge the time or the effort of winding a clock—and it takes me half an hour to get round the house on a Saturday morning—because I marvel how so little energy will be stored and spread over a week. For the same reason I do not despise the humble thirty-hour clock. Two hundred years ago or more some unimaginable gentleman or yokel gave this or a similar chain a single pull down every night, for years on end, and for each pull the pendulum was impelled to make some 85,000 swings. I like to add a few more hundred thousand to the total.

STRIKING

Most early clocks struck the hours. They were in their time expensive luxuries and one clock often had to serve for the whole of an ordinary house. Furthermore, bells were needed for the church which had an important role in the early development of clocks. There are two main forms of striking mechanism. The earlier, although it was never completely ousted by the other, is known as the 'countwheel' mechanism, and the later as the 'rack' system. They are both entirely reliable, but the countwheel system has the disadvantage that it must follow a cycle over twelve hours or whatever period the clock strikes. Once it has struck, say, ten, it cannot do so again for twelve hours; it can only strike eleven next time. Thus, if for any reason the clock fails to strike, or if you wish to push the hands round several hours to set them to time after it has stopped, you must operate the striking for each hour until the cycle corresponds to the time shown on the dial. With the rack mechanism, on the other hand, provided it is properly set up, the clock will always strike the number indicated by the hands.

You can tell which system is employed by a quick glance at

the works. If the system is the countwheel (also known as the 'locking plate') there will be, behind the back plate, or between the two plates in which the gearwheels run, a large blank wheel with a series of notches spread at graded intervals around its edge. Each notch is the same depth, and an arm falls into one of the notches when striking stops. So long as the arm cannot enter a notch the clock will strike; therefore the arc length of the projections between the notches is so arranged that each allows the required number of blows to be struck for each hour.

The rack mechanism is normally between the dial and the front plate of the movement. You will see a toothed rack, whose end falls onto a snail wheel, a sort of cam with steps of various sizes, one for each hour. The snail may be attached to the arbor of the hands, or it may be mounted separately. It turns forwards one step for each revolution of the minute hand. If a large number of blows is to be struck, the rack will fall to a low step on the snail and the striking train will revolve until an extension on one of its wheel arbors (which engages with the teeth on the rack) has raised the rack up to the top position again, when the train will be locked. If it is released again within the same hour, for whatever reason, the rack will still fall onto the same step of the snail and thus the clock will correctly strike the same number of hours again.

In both striking systems the actual hitting of the bell or gong is performed by a hammer whose tail is held, by a spring, in the path of pins spaced round a large wheel early in the train. This wheel, once the train is released, revolves and displaces the hammer tail, the hammer hits the bell, and the tail is returned by the spring to come into the path of the next pin on the wheel. The striking part of a clock is in effect a separate motor which is released at intervals by the time-piece position of the clock itself. It has, of course, no escapement; the speed of the blows struck depends on the size of a fan which turns at the end of the gear train to control its speed, on the strength of the spring or size of the weight which drives the wheels, and on the strength of the spring on the hammer which the pin-wheel has to displace.

We have now looked at the 'going' and 'striking' sides of clocks in general principles. In the next two chapters we shall survey the commoner types of clock in the light of these prin ciples. Whether or not a clock strikes is of some importance to the purchaser and, of course, to the ham as workman because there is a great deal more work in cleaning and setting up a striking clock (or chiming clock) than there is in working on a timepiece. The strike naturally influences design and the space available but it is entirely dependent on the time keeper, though usually it has a separate power supply. Therefore, we shall look at clocks in terms of whether they are spring- or weight-driven, because this is a more fundamental distinction than whether or not they happen to strike.

4

Weight-driven clocks

ESCAPEMENTS AND THE LONG-CASE CLOCK

The best-known of weight-driven clocks is the long-case clock. We will not call it a grandfather clock, because that is a Victorian name, the dignity of grandpa is perhaps not quite what it once was, and the ham likes to use the scientific terminology when he knows it. These clocks were made in increasing volume from about 1670 onwards until well into the nineteenth century. They have, of course, been made ever since, but more in imitation than in perfect continuation or development of the tradition.

There is not much doubt, I think, as to the reasons for the long reign of this type of clock. One of them we have already noted in passing—that the power of a falling weight is very much more constant over a prescribed period than is that of a spring and this was particularly true of the earlier period. There are ways of getting round the problem with the spring, but they are a complication and an expense. The great merit of the spring as a source of power is that it makes the clock portable; in consequence, and for other reasons, the proliferation of spring-driven clocks post-dates the real early advances in clock production and occurs when several portable clocks could be added to the static and dominating long-case clock in the hall or the living-room. Because the pendulum or balance-wheel of a spring-driven portable clock has a quick swing, a longer gear train is required and these clocks were, at least in the early stages, more expensive than long-case clocks and, for the collector, have continued so.

Another advantage of the long-case clock is its robustness. Apart from the very rare genuine miniatures only two or three feet high, the long-case clock is essentially a fixture and indeed, ideally, it should be screwed to the wall. This being so, and because a heavy pendulum is more reliable than a light one, and the weights required to keep a heavy pendulum swinging are correspondingly large, there is little merit in lightness or delicacy of construction. Of course there are long-case clocks ten feet and more in height and they take the principle to extremes in order not to be lost in some baronial hall. But the average long-case clock is at least six feet in height and ten inches wide in the trunk, though many early examples are slightly smaller and slimmer. If the clock has one weight, that will be of some 10lb; if it runs for eight days, there will be two weights, each somewhat heavier, and if it runs for a month or a year the weights naturally become sizable indeed. Since the balance of requirements was in favour of such a size, the dial was in proportion, being usually at least ten inches square, and there was no purpose to be served by having a diminutive movement. In consequence, the movements of these clocks are heavy and solid, their wheels large, their pivots (the slender end of the arbor on which the wheel turns) strong, and it often takes more than a century's filth and accumulated grease for them to grind to a halt. They are in essence highly reliable and reasonable time-keepers even in adverse conditions.

The uninitiated thinks of the 'grandfather' and its long pendulum as inseparable, and undoubtedly the supremacy of the long-case owes a good deal to the merits of the long pendulum and the simplicity and strength of the motor which is required to keep it going. But it is quite possible that the long-case did not owe its early popularity to the long pendulum so much as to the weights. It may well have originated as a case to protect and enclose weights hanging from early forms of domestic clocks, which might stand on a bracket or, more often, were suspended by a stirrup from a hook in the wall. To what extent, and exactly from when, it was so used we do not know and probably never will know, but certainly it was used before the long pendulum came in. Again, weight-driven clocks continued to be produced throughout the eighteenth century with no long case which is a convenience and a fashion rather than a vital necessity to the weight-driven movement.

These early clocks, without cases, were often of the type known as lantern clocks, from their resemblance to a brass lantern. Other earlier forms were made throughout of iron. They used, originally, a large and primitive balance wheel (being sometimes on the Continent just the two mid-spokes of such a wheel, known as a 'foliot', with little weights hung on them for regulation purposes and, later, a short 'bob' pendulum. Such clocks were driven by weights suspended from ropes; the rope was sometimes spliced into a loop and ran over a spiked pulley to stop it slipping. The escapement was the earliest known form, the 'verge' escapement, with a scape-wheel shaped like a crown with teeth sticking up from its edge (see Fig 2). Over this 'crown-wheel' swung a bar with angled pallets on it, if the clock employed a pendulum. If a wheel or foliot was employed, then the verge wheel stood vertically on its side and the pallets were also mounted vertically to engage with it.

For a long period this escapement was maligned for its inaccuracy, but, properly adjusted and with a short pendulum, it is now recognised to keep very reasonable time, though it is less happy with a long pendulum with its small swing and consequent small engagement of pallets and crown-wheel teeth. It has, however, the built-in disadvantage that the scape-wheel teeth are in almost uninterrupted contact with the pallets and it is consequently subject to the arch-enemy, friction. Many verge clocks were converted in the eighteenth and nineteenth centuries to other varieties of pendulum escapement. If you are interested in clocks for investment, you will naturally regard such conversions, which are usually detectable by signs of blocked or unblocked screw and pivot holes no longer used, with rather muted favour. If you are a ham proper, you will

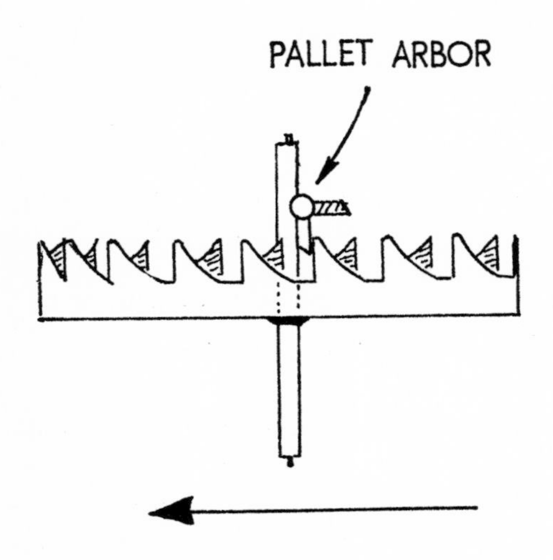

Fig 2 The verge escapement

mind less, and that is as well because you will save yourself being taken for a ride by a dealer who displays to you a 'genuine' verge clock, which in reality has not only been converted away from the verge, but has also been converted back again. Obviously, if you come across a genuine unaltered verge escapement, you will value it for its intrinsic interest, but do not be side-tracked into conceiving you have unbelievable good fortune, rather than merely good fortune, on your side; the escapement continued to be made throughout the eighteenth century, and in watches longer still.

One place where friction is especially unwanted is in the

mounting of the pendulum or balance-wheel. The balance-wheel cannot avoid it, though the trouble is minimised by specially shaping the very fine pivots of the wheel's arbor and, if possible, having them revolve on hard jewels or jewel substitutes which do not wear readily. The pendulum can avoid it, but in the earliest stages it did not. It was joined directly to the pallet staff and hung on knife edges; inevitably, there was trouble, and in due course the knife edges would wear flat. Subsequently the pendulum was not pivoted on the pallet arbor, but hung by itself from a silk suspension thread or supple spring. The latter form became, after the middle of the seventeenth century, the invariable practice for long-case and bracket clock pendulums. The pallet staff was connected to the pendulum, but was not directly joined to it, by a bent wire, forked or looped round the pendulum at the end, and known as the 'crutch'.

When we first considered the clock as motor and the pendulum as controller, it was said that the pendulum's merit was that the time it took to swing from one extremity to the other was independent of the size of the swing (the 'arc'). The matter is more complicated than that, and we will not go into the details here. Suffice it to say that there is error, though it is minute and, the smaller the arc, the smaller this error becomes. This was one theoretical reason for the demise of the verge—it was hardly the main practical reason because domestic clocks do not, and certainly did not, for the most part function with such precision that this error can be considered of very material import. It is, however, true that the verge escapement by its nature produces a large swing of the pendulum or balance wheel. This cannot increase its accuracy and does not render it very suitable to the long case and long pendulum which in the later seventeenth century were increasingly coming to be the vogue. The verge and a long pendulum require a wide trunk to the case, which was not in line with contemporary taste, and a great deal of power to drive them. Even then they do not function very satisfactorily. On the other hand, the verge had a longer existence in clocks of the so-called 'bracket' portable type because it was less disturbed by movement than were the newer precision escapements and went satisfactorily with the shorter pendulums.

Whilst the long-case clock did not have its origin in the housing of a long pendulum, and for all we know long cases may well have been common to house weights during the period (as much as five hundred years) when the verge was the only practical escapement, the long case and the long pendulum have been together for most of their lives, since about 1665. At this time a new escapement was invented which

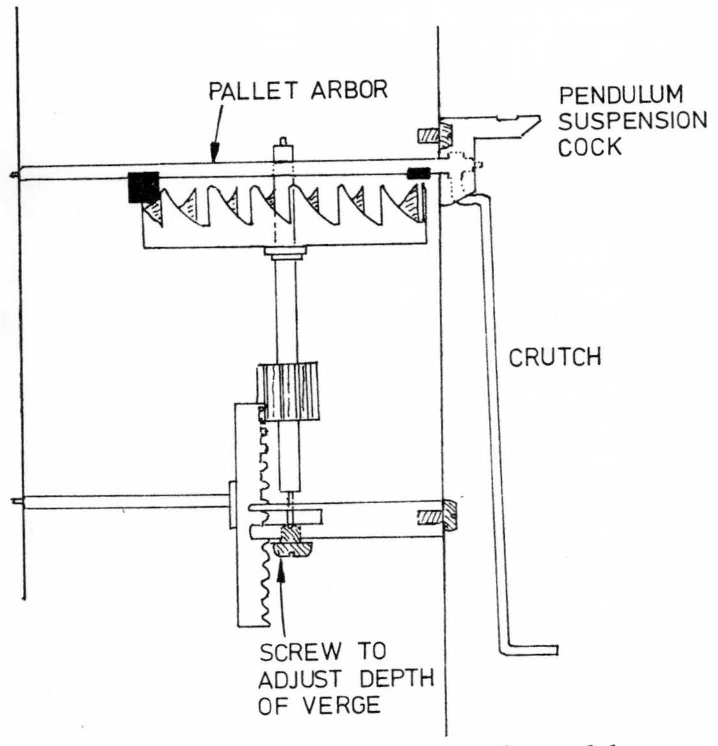

Fig 3 Verge and contrate wheel set-up for pendulum

revolutionised clock-making and time-keeping. It and developments from it are, and always have been since that time, the only pendulum escapements in general use. Others, in particular a highly exact and original escapement especially made for 'Big Ben' and used subsequently on timepieces of the highest class, have been used for special merits of simplicity in mass-production or of precision for single units. The new escapement is known as the anchor or 'recoil' because not only does the pallet bring the gear train to a sudden halt but it also, according to the precise shape of the anchor pallets, may cause it to turn momentarily backwards.

It will have been seen that the nature of the verge escapement demands that pallets and scapewheel were mounted at right-angles to each other with the wheel horizontal if a pendulum (which was more accurate and more easily regulated than the foliot without hairspring) was used (see Fig 3). This is not a defect in the escapement itself but it is an awkward feature of design in that, if the most economical use is to be made of the space, the rest of the gears must be mounted at right-angles to the crown-wheel also. This can only be done by means of a contrate gearwheel, which has technical difficulties of design and manufacture as well as practical difficulties of adjustment for reasonably efficient operation. One of the attractions of the new escapement may have been that this bugbear of the contrate wheel (which, however, reappears later in carriage clocks) was avoided. All the wheels could now operate in the same plane, between the same plates, with no necessity for sub-brackets to arrange for the change of angle, and the inefficient and wear-prone contrate wheel was eliminated.

If you compare the shape of the whole pallet forging in the recoil or anchor escapement (see Fig 4), you can see a resemblance to the verge pallets seen sideways on. There is a sense in which almost all escapements are developments of the verge system and this applies even to balance-wheel escapements.

Many of the same principles apply to all; there must be freedom in the action (termed 'drop'), or there will be wear in a very short time and irregularity due to excessive friction but, on the other hand, insofar as part of the escapement's function is to keep the pendulum or balance moving, all such drop

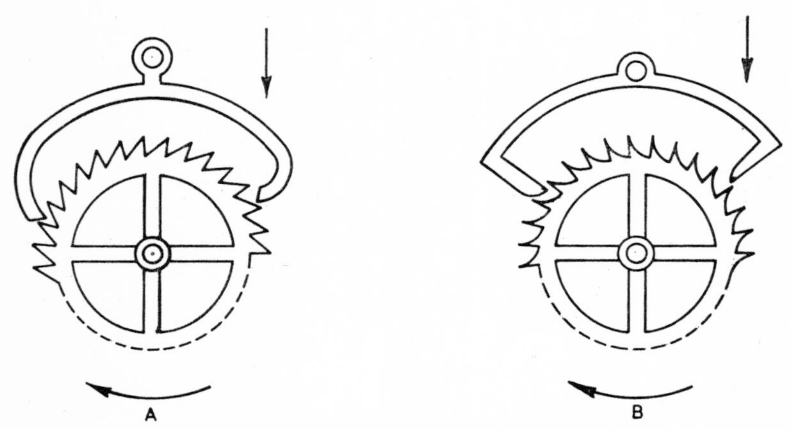

Fig 4 Forms of anchor escapement

represents lost energy in missed impulse, and therefore must be kept to the practicable minimum. There is a substantial body of theory on escapements, particularly on the angles of pallet faces to the scapewheel teeth and the distance of the pallet arbor from the scapewheel. My feeling is that the ham requires a knowledge of this data if he is to make an escapement from scratch, and he will find it helpful if he has to make a set of pallets. These are highly skilled jobs within the range of only a few hams, and you will learn most about escapements by watching correctly functioning examples and adjusting others until they work—that is, keep good time.

The pallets of a 'recoil' escapement are generally curved, and in the full form there is a corresponding curve on each of the scapewheel teeth. It is the continued sliding of the teeth on the long pallets after braking which brings about the recoil, which you can see in a backwards movement of a seconds hand if one is fitted. The recoil acts as a sort of brake on the pendulum which, if its arc became too large, could damage pallets and scapewheel. This escapement, and an early form with less recoil (see Fig 4A), was used on most, but by no means all, long-case clocks with the long pendulum. By 'long' is meant just over 39in, a 'seconds' pendulum which completes a swing in a second and facilitates a simple gear train. Longer pendulums were, of course, used in turret clocks for churches and the like. There was a limited fashion in very expensive clocks for a 5ft pendulum beating $1\frac{1}{4}$ seconds, the clock running for at least a month. In this case the pendulum bob swung in the base of the clock, rather than in the trunk as is normal. These clocks are rare and priced accordingly.

Although the seconds pendulum rapidly became all but universal after the invention of the anchor escapement, there is a good deal of variety in the shapes of the pallets used, though they are normally flat-faced rather than steeply curved and the wheel-teeth are straight rather than steeply curved in the back edge (the fully curved shapes being commoner for the short pendulums with relatively larger arcs in bracket clocks). As a matter of fact the shape is none too critical in practice, but the pallet shape and the drop need to be adjusted to keep the pendulum's arc as small as possible.

Within fifty years of its invention the anchor escapement was, for precision purposes in long-case clocks, theoretically superseded by a new invention—a recoilless escapement known as the 'dead-beat', whose debt to the anchor is clear from its appearance (see Fig 5). But the faithful recoil and its variations were never superseded—their sturdy quality of going when imperfectly made or badly worn (a quality with distinct appeal to the ham) ensured their survival. Forms of recoil escapement were and are still employed in perhaps the majority of domestic pendulum clocks.

The distinctive visual features of the dead-beat escapement

are the shortness of its pallet faces, the sharp undercut of the scapewheel teeth, and the positioning of the pallets. They are mounted much higher above the wheel than in the recoil escapements and they form a long-pointed inverted V rather than the rounded top of an anchor. This escapement is cap-

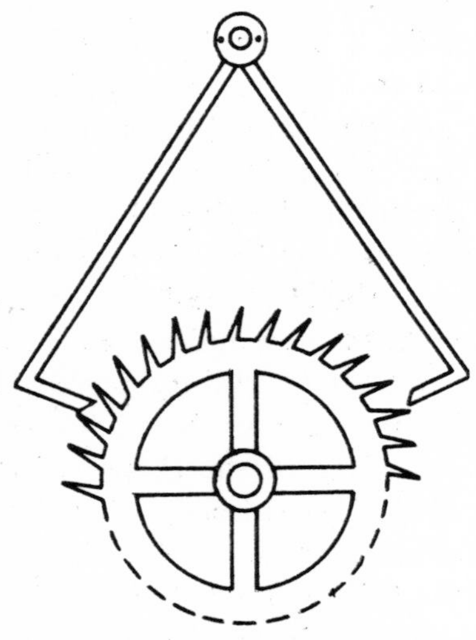

Fig 5 Dead-beat escapement

able of remarkable time-keeping when in good order but is profoundly unsatisfactory when in bad order. It was used in the more expensive clocks and its pallets were frequently jewelled to prevent wear. From its invention until the present day it has been the principal escapement used in long-case 'regulators'—clocks whose object is neither beauty nor even simple reading of the time on the dial but extremely accurate time-keeping for the purpose of setting a number of other clocks and adjusting their regulation. Such a clock is not, and was never, cheap. The regulator habitually eschews ornament and anything—like the need to set off a striking mechanism—which interferes with its single function, and it is always driven by a weight which gives the most constant power.

Page 67 Movement of thirty-hour clock (shown on page 50), which has a primitive anchor escapement and an early long pendulum

Page 68 Mid-eighteenth-century English bracket clock in style of c 1700-10

The pallets of the dead-beat escapement often embrace more teeth on the scapewheel than do those of anchor escapements and the arc of the pendulum is smaller. When the scapewheel teeth are intercepted by the pallet, they stop dead (hence the name) and the motor is not even momentarily thrown into reverse direction. Friction is reduced and precision is increased, but on the other hand the angles of the pallet faces are very critical, as is the amount of drop. Worn pallets will impair the time-keeping and soon stop the clock and, at least in the pure English form, the escapement is unhappy when the clock is moved and therefore not inherently suited to any but long-case or fixed clocks.

Forms do, however, exist which are used in table clocks, especially continental designs. This is the normal escapement for Vienna regulators (of which more later), which are fixed to the wall, and the type of French mantel clock which has a visible escapement in front of its dial uses a modification of the traditional dead-beat layout (see Fig 6). In this version the pallets proper are usually semi-circles of jewel-stone, often of

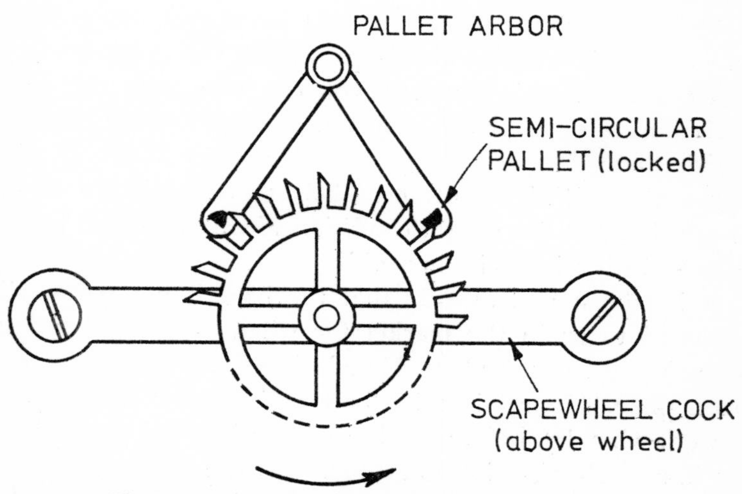

Fig 6 French visible type dead-beat escapement

a ruby red, mounted in brass with shellac. In cheaper versions, semi-circles of hard steel rod were used. One variation employs teeth cut away at the back and, in another form, the teeth are cut away to provide the impulse surface and plain pins (or occasionally gut strings, which are silent in action) are used as pallets. This latter variation is closely related to the pin-pallet lever escapement used with a balance-wheel in many cheaper watches and alarm clocks today. Again, it may be that the pallets are of fairly normal shape but the scape-wheel has a series of pins with bevelled edges instead of actual teeth. In yet another form, sometimes known as the 'half dead-beat', there is a compromise between the dead-beat and the recoil patterns; the pallets are shaped as for the recoil escapement proper, but the teeth are of straight dead-beat pattern and, as a result, there is some of the sturdiness of the recoil type and some of the accuracy of the true dead-beat escapement. This compromise is particularly common in long-case and bracket clocks of the later eighteenth century.

THE LONG-CASE CLOCK AS AN ANTIQUE

Enough has now been said of the types of escapement found in weight-driven clocks and in pendulum clocks generally. If you know the main patterns, you can cope with the many variations of detail which crop up. But what of the clocks in which these devices play so critical a part? What of their appearance? Which are most likely to be available in the rough, but not too rough, condition ideally required by the ham and what is it wise to avoid?

It has been said already that the ham is not necessarily attracted to or in the market for what, at the time of its manufacture, was the best. He admires florid engraving on a back plate. He rejoices in 'bolt and shutter maintaining power', whereby, before the key can be inserted to wind the clock, a mechanism has to arrange that the movement is not even for those few moments deprived of the power to measure the seconds and turn the hands. He would, of course, be delighted by a month long-case movement with calendar work and a five foot pendulum, and an original verge escapement with a bob pendulum would induce a state of ecstasy if it could be found. These and other bounties could not be resisted if they came his way and he had the necessary money in the bank. But he knows also that they rarely come his way and still more rarely has he the money required. Therefore, he is little less happy with the 'cottage' thirty-hour clock with one hand and a shattering strike on the hour, with a solid movement striking a jolly peal on eight bells, but not in its original case, or with a simple well-made timepiece in a tasteful small case of the nineteenth century.

Taste and sentiment have much to do with what attracts him, but he has some knowledge of what is and what is not typical of a period; he is no despiser of forgeries (most are highly interesting and many of exceptional quality for there is no point in forging rubbish) but he does not like having on his hands what he took to be genuine and becomes convinced at a later date is far from it. It may be an excellent clock which he would in full knowledge have strongly favoured, but in the circumstances he curses his ignorance and mourns the expense. Alas, we all do it; what the ham comes to learn is to appreciate a clock as far as possible for what it is—its mechanism, its appearance—regardless in some degree of what is or might be ascribed to it. It is true that its historical associations are bound to play some part in his estimation, but he tries to ensure, from painful experience, that it is not too large or too sensitive a part. Names are not everything and an inferior clock by a master may be less estimable than the product of an unknown and unnamed maker. Historical sentiment can play a part so long as you recognise it as such and that it costs money. As a ham you are not, after all, primarily interested in financial investment, although you should regard it as only prudent to break even whenever possible. The occasional monetary gain in fact or in prospect is rather an added pleasure, incidental to what you really value and enjoy.

Fig 7 Seventeenth-century long cases

The earliest long cases are attractively slim (see Fig 7). They consequently appear loftier than they usually are; in fact they are rarely much above six feet in height. The general impression is austere and restrained. Many of these cases are made of ebony or veneered in ebony, and from this overall blackness relief is given, in the bulk of the case, by rare and small details, like a brass door hinge or the plate of a lock, or brass feet and tops to columns on the top of the clock—the hood. The moulding around the base of the hood and at the foot of the clock is a reasonably reliable indication of date—it is almost invariably convex in the seventeenth century and generally concave, hollowed, thereafter. In the nineteenth century, of course, convex and other mouldings were used more freely, but even the novice is unlikely to confuse a nineteenth-century clock with one of the seventeenth century unless it is a deliberate and close imitation. Apart from ebony, these cases are made of oak, pine, deal, or walnut. Japanned lacquered cases date from about 1710 but are commonest on clocks of the early mid-eighteenth century. Before that time, decoration was in the form of marquetry, often employing coloured wood such as olive and box. Mahogany was known, and very occasionally used, but it did not become at all fashionable for clocks until well into the eighteenth century. Early London-made clocks from about 1685 to 1700 tend to have delicate 'barley-sugar' turned pillars on each side of the hood, whose top is for the most part plain and square, although some tops have, or had, carved and fretted crestings on the front or on all sides. The London clock is also distinguishable by the fashion for glass panels in the side of the hood.

The early history of the long-case clock is hard to determine, and the probability is that it never will be fully determined, since it is plain that practice varied very widely up and down the country. Fashions often started in London, but they tended to survive there, and still more so in the provinces, long after the height of the fashion had gone. An instance is the thirty-hour clock already mentioned. There are those who hold that it is the earliest form of long-case clock, and certainly there are examples by the earliest known makers. But the same authorities will not deny, indeed they assert, that such clocks continued to be made, presumably for their simplicity and relative cheapness, throughout the eighteenth century especially in country areas, and that there are clocks from the earliest times which run for more than thirty hours. The same applies to the lantern clock, whether spring- or weight-driven. Therefore, while it is likely, if other features (like turned columns, a square dial of smallish size) are present, that a thirty-hour clock with a single hand and a movement designed like those of early lantern clocks and domestic wall-clocks is in fact an old one, there is also the possibility that it may be up to a century younger than first impressions suggest. The ham who finds such a clock where the maker is not named and the date not fairly certain, will no doubt consult the illustrations in standard works and the examples in museums and come to the best decision he can. From the first, the eight-day long-case was at a premium. Therefore, though most thirty-hour clocks have no winding squares or holes for these in the dial, the occasional thirty-hour clock is found with holes and dummy winding squares. The presence of rope or chain to a single weight, however, gives the game away on inspection.

The earliest long-case clocks have hoods which slide up rather than hinged doors for access to the dial. This may be because so many were wound by pulling up the weight rather than by inserting a key into the front, and because they tended to have countwheel striking systems, so that if the clock had to be adjusted by more than a few minutes the hood would have to be removed to adjust the cycle of the striking. We know, however, that the upward sliding hood continued to be used (though it remains an early feature) after clocks had minute hands, ran for eight days and even after rack-striking became fairly general (say by 1705). Though there was a catch to secure the raised hood for winding, the upward sliding hood could obviously be a source of inconvenience, especially on a change of ownership to a place with low ceilings, hence a great many such clocks have been converted to slide forward (as many original hoods do) or have been fitted with a hinged door. Practice varied so widely that a fixed date simply cannot be given for such changes. Again, we know that there was a fashion for engraving rings round the keyholes of front-wound clocks, no doubt because of the unsightly marks which keys can make on a dial after a few years, but we cannot say with any certainty just when, between about 1710 and 1725, this fashion started and sometimes rings were for sound reasons added to clocks dating from long before the embellishment was fashionable. The same may be said of the 'lenticle', the glass-covered hole in a long-case clock's trunk door which appeared in about 1685 and which had the clear purpose of showing whether or not a clock without a seconds hand had stopped. It continued, and was added to earlier 'solid' fronts, long after 1700.

The study of dials and hands is a science in itself and much has been written about it. The dials of the seventeenth century are small—they rarely exceed eleven inches, and ten inches was the general fashion in the last years of the century. They are invariably square in appearance and nearly so in measurement. They are made up—as are nearly all dials until the mid-eighteenth century and many thereafter—of three main components, the ring with the figures (known as the 'chapter ring'), the corner pieces ('spandrels'), and the basis of the whole thing, a heavy brass plate. The earlier the dial, as a general rule, the simpler the spandrels and the greater the relief of their chasing. Later ones were mass-produced and often applied straight from the casting without elaborate finishing-work, but are more complex in pattern and flatter than the hand-finished article. Spandrels are nearly always screwed on and, if one were mislaid, no doubt the whole set would be replaced by a later pattern. The centre of an early dial is usually matted—covered with minute dots from a punch—whilst in the eighteenth century the centre is often elaborately engraved.

The design of the chapter ring varies according to date and also according to the nature of the movement but these two factors do not always exactly correspond. The single-handed clock has a ring with four divisions between each of the hour marks. When minute hands became usual (which was not long after the adoption of the anchor escapement and seconds pendulum) the minute divisions were of course used, so that then there are five divisions between each hour. But, as always, we are faced by the fact that fashions sometimes outlive their purpose, conversions are made, and stocks have to be used up, so that it does not follow (though of course it is the usual presumption) that a 'quarter' ring where there is a minute hand represents a 'marriage' of parts which were not always together. As a transition (until about 1750), you will also find a sort of dual-purpose dial, where the hour ring has four quarter divisions and the characteristic branch motif at the half hours, but there is also a full and numbered minute ring outside. In this connection also, it is worth repeating that single-handed clocks continued to be made for the greater part of the eighteenth century and their dials may or may not have minute divisions.

The earliest chapter rings were narrow, and the matted centre space correspondingly large. There seems to have been a contrary fashion in the early eighteenth century, especially in one-handed clocks, for very heavy and deep chapter rings, but soon the proportions were evened out by the larger dials. Minute figures, if given, are in arabic style and between the two circles of the minute ring until about 1700. Thereafter they are usually outside the rings and often, to modern taste, grotesquely large and obtrusive. No doubt, for a while, the keeping of time to the minute was a luxury and the divisions unfamiliar at a quick glance. The maker's name is usually on the chapter ring, or above it in the arch when there is one.

A name on the dial plate below the ring is, subject to the context, likely to indicate an early date.

In the matter of dial shape there is a fairly well-marked division between the square type and the type with the rounded arch on top, which became fashionable around 1720. In some instances the arched top was added to existing dials as soon as it came into vogue, and perhaps before the clock left the workshop (it is a pity that we know so little of the stock-movements of early manufacturers). Such an addition may be a genuine curiosity rather than a botched repair or imitation. The space provided by the arch was soon used, particularly in bracket clocks but also in long-case clocks, for subsidiary dials, commonly one being for regulating the pendulum from the front without stopping it, and the other to switch the striking system on and off. The space might be occupied merely by the ubiquitous medallion inscribed with 'tempus fugit', or some other dictum hallowed or curious, by the maker's name or, less generally, by a set-up showing the times of the tides or the phases of the moon. The arched dial was accompanied by modifications in the style of the case, which became generally more massive and a great deal higher. It was met also with the gradual introduction of more curves into the lines; first the fascia surrounding the top of the dial was curved, then the top of the door in the trunk, and, later still, towards the end of the century, the whole top of the case was made to follow the curve of the dial.

Hands follow the general rule of spandrels, in that the simplest are the earliest but it is impossible to survey the subject in detail here. They are a much more accurate indication of date than are spandrels, because the vast majority were filed up by hand until the late eighteenth century when they started to be pressed out in batches. The earliest hour hands have a simple spade-like tip, sometimes almost heart-shaped, later ones having a longer point and elaborate patterning. The earliest minute hands are straight and thin for most of their length with an offset twist or curl close to the boss. It is not difficult to recognise seventeenth-century hands of this type, but thereafter the varieties become more numerous and if you are uncertain you would be well-advised to consult the extended discussions of hands in Britten's *Old Clocks and Watches* or Cescinsky and Webster's *English Domestic*

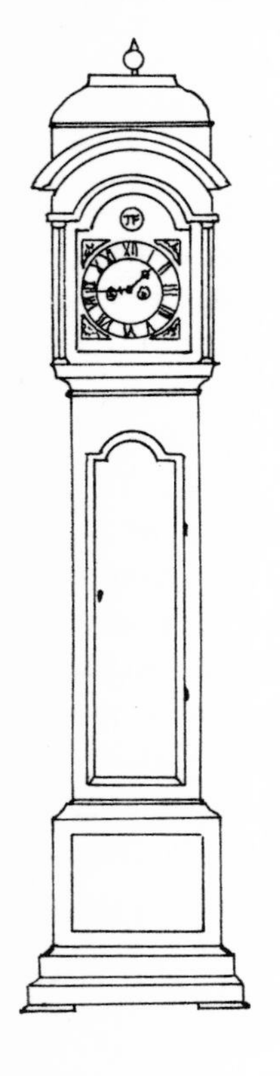

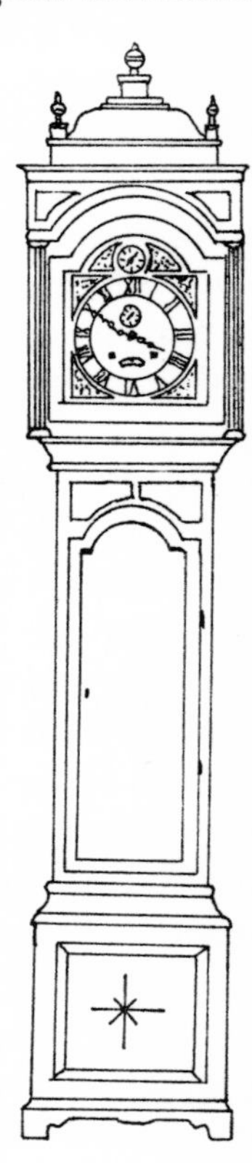

ABOUT 1720 ABOUT 1750

Fig 8 Eighteenth-century long cases

Clocks. It is worth remembering that, in general, the more alike are the hour hand and the minute hand, the later they are likely to be.

After the middle of the eighteenth century, marked changes occur in long-case clocks, though the older patterns continued to be made (see Fig 8). Those made outside London tend to become very much broader in the trunk and often have a somewhat squat appearance (see Fig 9). The thirty-hour clock virtually dies out. Ornament is prominent and elaborate crests are common but movements, being more or less mass-produced and assembled by the provincial maker, are simpler and plainer. The top of the hood may revert to what is in fact the earliest shape—a classical pediment, although usually adorned now with an eagle or a ball on top and at the corners —or it might be completely bowed. Doors are usually shaped at the top, sometimes curved, or with the corners mitred off. The base and hood become very much larger in proportion to the clock as a whole and the trunk, in extreme examples, is hardly more than a waist albeit a plump one. Dials vary a great deal in shape and round dials begin to appear in the last twenty years of the century. Another characteristic is the finish of the dial; spandrels are flat castings or disappear altogether, the chapter ring as such may not be imposed on the dial plate, the whole dial being silvered and the rings and figures engraved into it. Alternatively the dial may be of painted enamel, often with simple indications of tides and moon phases above or below the central arbor, or with a rocking device, such as a ship, linked to the pallet arbor. The seconds hand was used throughout the century but becomes much more general later, as does the simplest calendar work —an aperture (usually straight in London and tending to be curved elsewhere) showing the dates from 1–31, but not, in the standard pieces, the day or the month.

These later long-case clocks have until recently been somewhat underrated. It is true that many were produced with little attention to finish well into the nineteenth century, and true also that their broad, four-square appearance, especially in pieces by northern makers, can be less dignified than either the severe simplicity of the earliest types or the lofty and monumental proportions of many of those in the middle period. But the fact remains that they are well made, reliable,

Fig 9 Late provincial long case

and more likely than older clocks to be in original condition (this applies particularly to the earlier London clocks, which have long been sought after and tampered with).

OTHER WEIGHT-DRIVEN CLOCKS

Rather similar questions of taste and value arise with so-called Act of Parliament clocks, a type which is rather a clown among the weight-driven breed (see Fig 10). These clocks, which date mainly from the 1790s (though they are probably less closely associated with the Act placing a tax on clocks than their name suggests, and outlived it by many years) are intended for showing the time in public places, such as public houses. Their distinctive feature is an enormous face, often two feet or more in diameter, and either round or octagonal. Below this is a relatively small trunk for the pendulum and weights. Sometimes the pendulum bob is visible through a glass aperture. The dials vary considerably, but the emphasis is clearly on showing the time across a large room or hall. One of the commoner styles has large Roman figures for the hours and arabic figures, not much smaller, for the minutes. The figures are painted directly onto the wooden front of the case, which is usually dark brown or ebonised—at this stage there were problems in enamelling a dial surface of such a size. In general, while there are rare exceptions, there is no glass on these clocks and they do not strike.

Finally among the range of weight-driven clocks commonly seen comes the large and somewhat miscellaneous class of 'Vienna regulators'. The name should, perhaps, be restricted to the finest examples, with full dead-beat escapements, regulator-type dials (that is, separate and not concentric rings for minutes and seconds), and no striking mechanism. The other extreme is the mass-produced pendulum clock, spring-driven, striking, often rather discordantly, on a gong, and usually with some form of recoil escapement. In between there is a whole range, including fine clocks, weight-driven, with dead-beat escapements but having striking trains.

The Vienna regulator is really a weight-driven clock of quality for hanging on the wall. It represents the final stage in the development of weight-driven clocks, having a refined escapement, a relatively heavy pendulum, and driven by surprisingly light weights. Most often the pendulum rods are of wood with brass-cased round bobs; superior examples have elaborate compensating pendulums comprising a series of rods with different rates of expansion to compensate for changing temperatures, or special forms of bob for the same purpose. The pendulums beat two-thirds of a second, and in consequence there are examples with only forty divisions on the seconds dial or, more usually, there are sixty divisions and the

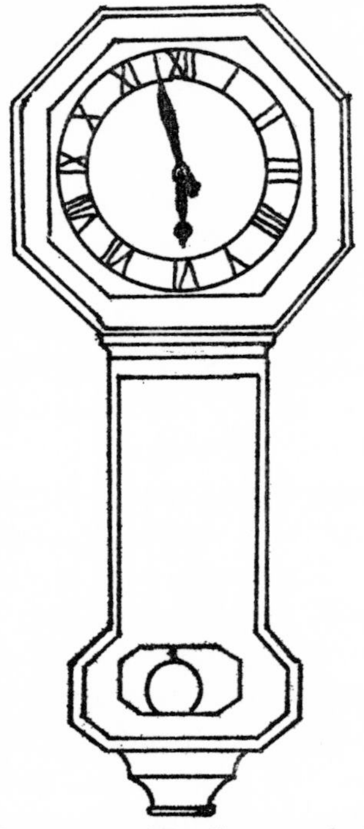

Fig 10 Act of Parliament clock

hand does not line up with all of them. The weights are cast in brass cases and suspended from well-finished brass pulleys which run on fine gut, never on chains. The pendulum is mounted separately behind the movement on a stout casting. The escapement is a precision device with the informed owner in mind; the pallets are adjustable and reversible and the pendulum's position relative to the pallets, (the 'beat' already mentioned) is adjusted by screws on the crutch. The rarer Vienna regulators run for a month, but the majority go for eight days. They were made in the last quarter of the nineteenth century, and the more excellently made ones are valuable. There are still, however, a great many bargains to be had in Vienna regulators by the ham who looks around

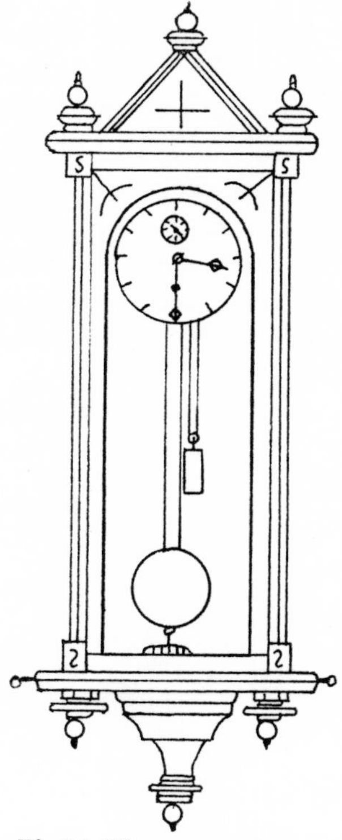

Fig 11 Vienna regulator

and is not taken in by inferior imitations. A good striking version is a clock well worth having and can still be found reasonably cheaply. Chiming clocks are very rare—the three weights required are not a practical arrangement for a wall clock and to operate a chime as well as a strike is asking much of a movement of this type.

THE STRUCTURE OF WEIGHT-DRIVEN MOVEMENTS

There are two main forms of weight-driven movement (see Fig 12). The one, descending directly from the earlier lantern-clocks, is assembled in a frame with four iron upright posts riveted into brass above and below. Midway between the side posts are three strips of heavy brass, a sort of skeletal movement plate, in which the wheels run. These strips are usually slotted into the top and bottom and held by iron wedges. The 'going' part of the movement is to the front of the clock and the striking mechanism and wheels are behind. These movements, known familiarly as 'birdcage' movements from their appearance, employ the countwheel striking system and the

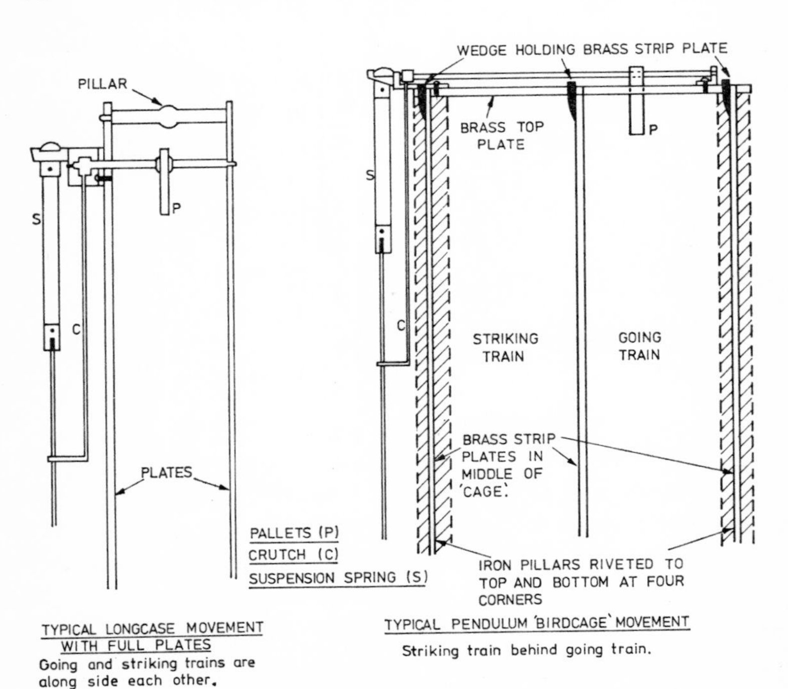

Fig 12 Layout of long-case movements

countwheel itself is most often pinned or clipped onto the back 'plate' outside the movement, although it is sometimes placed inside.

There is a predisposition to regard this type of movement as 'early'. Very often they are, and very often they are not, depending on what you mean by early; they continue, often with only an hour hand, until the middle of the eighteenth century, but certainly they are likely to be at least 200 years old.

This form of construction was rapidly superseded in the late seventeenth century for the vast majority of clocks with the invention of the anchor escapement, the use of the long pendulum and the proliferation of smaller and more complicated 'bracket' clocks driven by springs. The general practice became to mount the wheels between solid vertical plates, usually two (but occasionally three), joined at the four corners by turned brass pillars. The plates were rarely engraved for long-case clocks since they would not be visible inside the case. The shape of the pillars employed gives a rough indication of the age of the movement (see Fig 13). The earliest ones were curved, known as 'baluster' pillars (stemming from

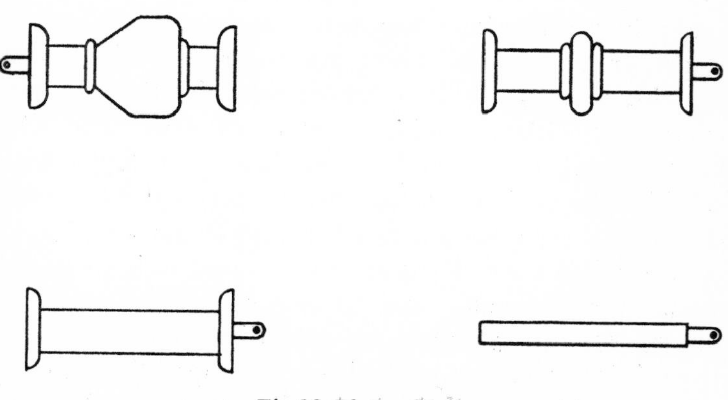

Fig 13 Plate pillars

the oldest, horizontal, movements), those of the middle period (from about 1680 to 1740) were straight but with a round hump in the middle, and the later ones were plain and cylindrical, but many exceptions to these tendencies are found. The pillars are riveted or screwed into one of the plates; practice varies as to whether this is the front or the back plate but the front is more usual in English clocks. The other end is most commonly secured by a pin, but in the oldest movements a little clip latch swings round into a slot in the end of the pillar to secure it. The 'striking' and 'going' trains are arranged side by side, and general but not universal practice is for the striking mechanism to be on the left when seen from the front. In outline these principles of construction and layout have not changed and they were applied equally in mantel clocks, where they are followed to the present day.

BUYING A WEIGHT-DRIVEN CLOCK

What, in all this variety, is likely to be found, and what is to be avoided? Basically, the risks in buying long-case and other weight-driven clocks are relatively small, and, unless one insists on well-known names and superlative pieces of furniture, extravagances in their day and still more so now, it is still possible to find clocks of almost all vintages available, though their prices have begun to rise rather steeply in the last two or three years. There are few parts which are not repairable and a great many which can, with time, be made satisfactorily without sophisticated equipment. The wheel with a missing tooth can have another tooth dovetailed in (but this damage is far commoner on spring-driven clocks). Pendulums can be bought new from the materials shop, and so can weights. Pendulums are largely, and weights to some extent, interchangeable and in any case quite reasonable replacements can be made up specially. Ropes can be re-spliced, it is not an easy job and new soft rope is hard to come by, but a yacht or small ship's chandler has an ample choice.

However, chains, and pulleys to fit them, can be bought. Reproduction hands are available and can be tastefully finished off with a file, or you can make your own in one of the older patterns, but it is a long job. The commonest missing part of the thirty-hour clock, namely the external countwheel, can be made fairly easily, as can the simple gearing which goes with it. Worn escapements can be repaired, and missing pallets can be replaced with a standard forging obtainable from the material dealers and finished to size, but the old anchors have to be replaced or repaired by hand. Bells, often missing, are also available. Gut or twisted steel wire for the weights of an eight-day clock is available, as is the finer size for the Vienna type of clock. Holes can be bushed without difficulty, and you can often replace broken pivots; if you find this is not possible, the turning of a new arbor or pivot is not an expensive job to have done. The silvering and figuring of chapter rings can be re-done quite satisfactorily. Reproduction spandrels are available, although it is usually a case of replacing with a new set which may not be quite the same as the originals in style. Even a whole dial can be made up from a bought chapter ring and spandrels and a sheet of heavy gauge brass with rods screwed into it for the dial 'feet'. It may not be original, but there is no reason why it should not be both tasteful and suitable in appearance and thus give the old works a chance to run again. Damaged cases can be repaired. The services of a cabinet-maker can be employed if complicated marquetry needs to be restored and the job seems worthwhile. When a case is missing a reasonable replacement can be made up and employed until a more genuine case comes your way. Clock glasses of all kinds are available from materials shops if you cannot cut the less simple shapes yourself, and a do-it-yourself shop should do it for you if you provide a paper template of the required shape.

Nonetheless, there can be problems. Serious wear in the pinions throughout a movement may seem to be a common one. The remaking of the whole set is not lightly to be undertaken and it is an expensive job to have done. Fortunately it is not likely to be necessary or desirable; one or two of the smaller pinions may have to be replaced, but you want this movement going rather than a new one instead of it and a little looseness in the later gears of a train, though causing noise especially in the striking mechanism, is not likely to stop the clock. A very fine pair of hands, one of which is broken, is best taken to an optician's or a really good jeweller's for hard silver-soldering; the minute hand of a long-case clock takes a considerable strain when it is set to time and even the softer silver solders do not always make a strong enough job. The conversion of a rope-driven clock to chain drive, if your conscience permits it, is a job which cannot usually be done without the use of a lathe or someone to use it for you, although the turning required is of the simplest. A clock chiming a tune employs a pinned barrel like that in a musical box; odd pins can be replaced, but if a complete barrel is missing the designing and making of a new one is an expensive business and probably outside your scope but, of course, such a clock is worth more on the market and the repair may well be worth having done. Again, the grooved barrel around which the gut or wire of eight-day clocks runs can hardly be made in the average workshop, but it may be possible to work on a replacement from another movement. Finally, there is the problem of the enamelled dial which has had a hard life. Plain white can be touched up, but the sometimes elaborate pictorial illustration on these dials—which can strike one as hideous or charming as much for reasons of personal taste as for their intrinsic qualities—is not an easy job. Even the figures can be very difficult, because they are not engraved, as are those on the silvered chapter ring, and so offer no guide to pen or brush. If you are taken with a particular painted dial and it really is in a bad way, make enquiries for an artist before buying it, or leave it alone unless you have had some successful experience in this line or know of one of the few firms who undertake such work. If the particular dial is not a main attraction, then others are not too difficult to acquire (possibly together with a movement for which you have no immediate use) either in brass or in enamel. Be warned, however; genuine old brass dials are in demand for just the purpose which you have in mind, and they are becoming expensive.

Many of these snags leap to the eye, but the eye may be so taken with the appearance of the desired object that it fails to notice them. A simple mechanical inspection will detect the more latent troubles. Remove the pendulum and the pallets, if these are there, and let the going train run. It is safer to detach the weights and pull on the gut or, in the most derelict instances, push round the largest wheel with a finger. The wheels should run reasonably smoothly. A missing tooth or pinion leaf will be detected by odd bumps in the running, by the jump forward of the minute hand, or it can be a reason for being unable to move the train at all. Be prepared for stiffness in a rusty and dirty movement. If there is no dial, make sure that there are two wheels on the central arbor to which the hands will be fitted (or one wheel, usually loose, in a single-handed clock); you can probably cope with missing hands and a missing dial, but missing motion wheels are a serious drawback. When you have done this rough check on the going side, apply the same tests to the striking system after freeing it by turning the hands and holding the rack or count-wheel lifting pieces out of the way. The clock need not strike correctly to satisfy you, but at least the train should run. If you are uncertain of your power to make the thing go, replace the pallets and crutch, but not the pendulum, and again cause the going wheels to move; an escapement not too far out of adjustment should cause the crutch to swing to and fro—however, it need not do so reliably or evenly at this stage. If you can persuade a weight-driven movement to function this far manually, it is most unlikely that, however limited your equipment, you will not, in due course, be able to set it up adequately at home.

Subject to these hints, the matter is one for everyone to decide for himself. If you never buy anything but a perfectly running clock, you will lose one of the essential ingredients of the pleasure of being a ham, the chance of gaining experience, and also spend a great deal of money. On the other hand, you would be foolish to pick up an incomplete wreck of a movement that can only be made to go by a course of replacement so far-reaching that it scarcely remains the same clock at all. You have always to estimate what you can do and what you can afford and balance them. Then you take the plunge. Sometimes it is a misadventure and takes time to forget but, in any event, you have usually acquired a valuable store of parts and material. More often, however, you have a great many hours of pleasure in front of you and an object of some beauty and antiquity constantly in your company thereafter.

5

Spring-driven clocks

THE SPRING AS A SOURCE OF POWER

The great difficulty with the spring is that its strength does not remain constant, either through its working life or from day to day. Its great merit is that it is the means of making clocks portable. We will consider first the means of getting round the difficulty, then the common types of clock and their escapements which exploit its merit.

The principal ways of levelling out the power supply are the stackfreed, the fusee, and the going barrel. The stackfreed hardly concerns the average ham. He might occasionally find it in an ancient table clock but the chances are remote. Suffice it to say that the stackfreed consisted of a stout spring, bearing on a cam fixed directly on to a wheel arranged to rotate once during the unwinding of the clock's spring. At full wind, the spring bore heavily on the cam, and when the clock was well run down the spring barely pressed on the cam at all.

The stackfreed's interest is historical only—it was something of a gesture for spring-driven clocks where space was limited—but the idea that the fully wound spring must be counter-balanced by a disadvantage, which must be reduced as the spring runs down, is in simple form that of the more important fusee (see Fig 14). The fusee is, in principle, a continuously variable gear; when the spring is fully wound it turns what is, in effect, a small wheel, which in turn has to operate the full train of gears and, as the spring unwinds, it operates the train through progressively larger gears until eventually it is employing the fullest diameter of the fusee barrel. In actual fact the fusee is not a series of wheels and the spring barrel does not connect to it by gear teeth. It is a brass cone with a groove spiralled round the circumference, and a toothed gearwheel, the first of the train, as its base, to which it is connected by a pawl and ratchet. The spring itself is coiled into a large plain barrel which rides on a separate arbor and it is connected to the fusee, and thus drives the train, by a chain, gut line or wire. When the clock is wound this gut line is coiled onto the fusee. As the spring unwinds on its arbor which is fixed, the barrel turns and draws the gut off the fusee and on to the barrel, which it can only do by pulling the fusee and the whole gear train round. It sounds complicated, but in practice it is not so. All you have to remember is that when the clock is fully wound all the gut is on the fusee (where it presses up a finger which stops the clock being wound any further) and pulling on the narrowest part; therefore the 'gear' of the fusee only allows the mainspring to work at a mechanical disadvantage, which is decreased as the spring unwinds. Precisely when this device was invented is not known. It existed on the Continent in the late fifteenth century and is by far the most efficient device for its purpose which played a part of incalculable importance in the development of spring-driven clocks, and, at least in England, will be found in almost any mantel clock of high quality until the mid-nineteenth century and indeed thereafter.

The practical alternative, and in the modern period of reliable steel and cheaper clocks, the general one, is not a separate device but simply a deliberate restriction of the mainspring. A clock spring running freely will unwind, subject to what it has to do, until the clock stops through lack of power or the spring runs into an obstruction such as one of the pillars which hold the plates together. A spring in a barrel can unwind only as far as the size of the barrel permits. Thus, though the clock will run for only part of the period over which an unenclosed spring might manage to drive it, if the spring is in a barrel the running time will be the period of tension when the spring is near to its full power. Thereafter,

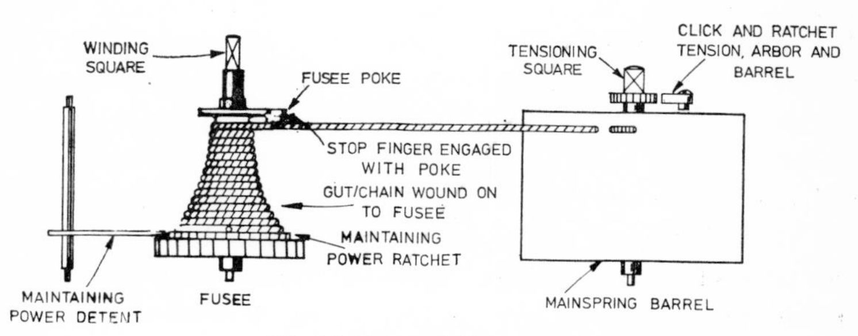

Fig 14 Fusee fully wound

the spring will be unable to expand any further and the clock will stop. It is a simple arrangement—there is no need for a separate barrel and 'loading' device, and none of the gut or wire which in a fusee clock will always put the movement at risk of disaster should it break when the spring is fully wound. The going barrel is simply extended and toothed at the bottom so that it becomes the first wheel of the gear train.

THE SPRING-DRIVEN CLOCK AND ITS DEVELOPMENT

Spring-driven clocks of the 'classical' period of English clockmaking—that is from the late seventeenth to the early nineteenth centuries—mainly have pendulums although earlier clocks, like their weight-driven counterparts, had crude balance-wheels or foliots. There is little to be said of their escapements which has not been outlined in connection with weight-driven clocks, save that the verge escapement persisted very much longer, even into the early nineteenth century. The reasons for this are subjects for conjecture and some have already been mentioned. Possibly, it was found that the verge with its large pendulum arc, though it might not keep the best time, was less likely to be stopped if the clock were moved. Perhaps the light 'bob' pendulum (which did not last as long, but outlived by half a century its demise in long-case clocks) had similar advantages. Certainly the verge escapement is robust compared with many later forms and it will run after a fashion even if the clock is not set perfectly level. Bracket clocks seem, in general, to have received more elaborate care and craftsmanship than long-case clocks by the same makers and there is reason to think that the so-called 'bracket clock' was very often a bedroom clock—a luxury rather than the prime time-keeper of an establishment. For whatever main reason, the fact is that the verge and short pendulum seemed to go well together and continued to be made long after the seconds pendulum in long-case clocks had been fitted with more refined escapements. The anchor escapement, in its various forms, was not preferred in spring-driven clocks until about 1740. The evidence is unclear, however, since a large number of verge clocks, which no doubt performed perfectly adequately, were converted to the anchor, perhaps partly so that the wider and heavier pendulum bobs of the time could be accommodated in the same case which had previously housed a 'bob' pendulum and verge with their larger arc.

It is obvious that, even if it has the edge in time-keeping, the pendulum is inherently unsuitable for portable clocks. The fact was perhaps partly obscured by the trend towards larger and ever more elaborate spring-driven clocks in the eighteenth century, or it may be that the fact itself contributed to the trend. But certainly throughout that period the search went on for an efficient and really portable form of controller and escapement. Such an escapement was in fact invented in 1770. Yet this, the lever escapement, which for all practical purposes has been the standard escapement for watches and small clocks for the last 100 years, was not really exploited until about 1830, although certain more fragile and less satisfactory escapements (such as the cylinder) had been known and used in watches very much earlier. The reasons for this are debatable. One may have been the slight defect in design whereby the balance and lever of the early escapements could easily be jolted out of position and cause the clock or watch to stop.

It is not difficult to see in the lever and cylinder escapements a fundamental similarity with the escapements of pendulum clocks (see Fig 1). There are two pallets spanning teeth on a basically ratchet-toothed scapewheel and they alternately lock the scapewheel and provide an impulse to keep the controlling balance-wheel and hairspring oscillating. The resemblance in the cylinder escapement is slightly less apparent, but, effectively, the two edges of the cylinder amount to two pallets embracing one tooth of the scapewheel. The body of the lever may be equated with the crutch of a pendulum clock and through it the impulse is transmitted to the balance-wheel, usually through a pin on the wheel arbor, with which the forked tip of the lever engages.

You will have noticed in the dead-beat escapement that its great merit is in the relative absence of friction, the shortness of the pallet faces, and the consequent independence of the pendulum. This is of importance to the time-keeping of the pendulum, and still more obviously is it vital in the case of a balance-wheel if it is not to grind to a halt. Lever escapements are classed as 'detached' escapements; the controller is in contact with the scapewheel only momentarily through the pallets and lever. But with the cylinder escapement it is quite the opposite; the balance-wheel is in constant contact with the scapewheel and highly subject to any variation in the power of the main spring. This effect on time-keeping, and the fragility and liability to wear, are the main reasons for the final superseding of the cylinder by the lever escapement early in the present century.

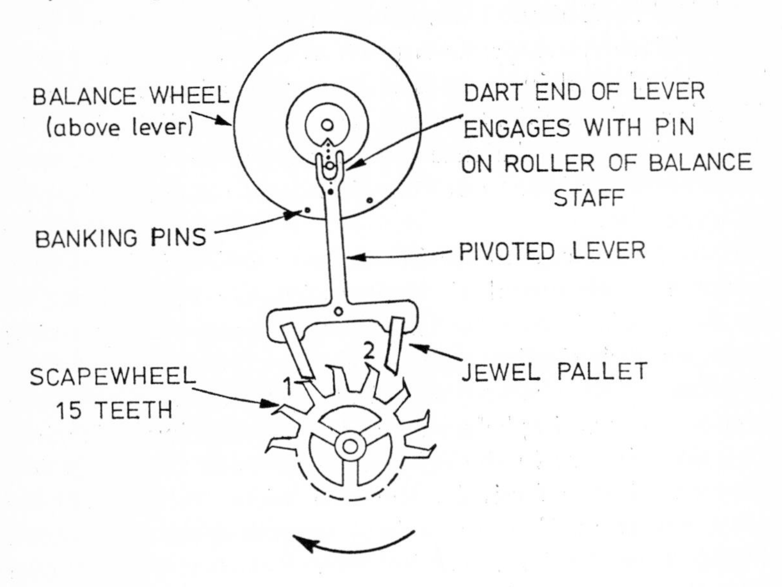

Fig 15 Club tooth lever escapement

The lever escapement varies very widely in form—the form illustrated (see Fig 15) is the most general modern version, the form which is available in high-quality units at reasonable prices for replacement purposes. Many earlier versions place the pallets at right-angles to the lever to which they are fitted, not being one part (see Fig 1). These have scapewheels of brass rather than polished steel, and the teeth are pure rachet teeth rather than hooked 'club teeth'. There is a modern version of the escapement where the pallets are steel pins and the scapewheel teeth cut back very much as those on the 'pin' forms of dead-beat pendulum escapement in French clocks. The detail of connection with the balance arbor (the 'balance staff') also varies considerably. One early form, still occasionally met in old watches, was very far from detached at this end—the balance staff was formed into a pinion which engaged continuously with the end of the lever which was formed into a toothed rack.

It is not possible, nor would it be profitable, to list and fully illustrate all the varieties of the lever escapement. You will quite often come across what, by modern standards, are oddities. You will find helical rather than flat hairsprings (used in the attempt to gain isochronism), scapewheels with upper and lower sets of teeth, pallets and levers of different shapes and sizes. If they are in fair order you will not have any difficulty in eventually getting them going, though it may be necessary to consult one of the standard works (such as W. J. Gazeley's *Clock and Watch Escapements*) to check on pallet angles and tooth shapes. If you can get an old form of escapement going, you have a great sense of achievement and a distinct novelty in your collection. But most often these escapements are severely damaged or incomplete. To have such a piece restored or a new escapement made is an expensive business and will not be worthwhile save for the investor in really expensive clocks. A better course is to have the clock working with a respectable modern escapement and to hope that in due course an escapement of the original type, perfect or repairable, may eventually turn up, when it can be fitted. In this field, you need to be ever-hopeful.

THE SPRING-DRIVEN CLOCK AS AN ANTIQUE— BRACKET CLOCKS

What, now, of the clocks to which these devices were fitted? Only rarely does the term 'bracket clock' actually mean a clock which has or had at one time its own bracket for fixing to the wall. It is generally used of mantel and table clocks of the 'classical' period and rather later—sumptuous pieces almost always with pendulums and fusees and with dials constructed much as are those of long-case clocks. Others, especially those wishing to sell less happy modern chiming clocks with an element of the reproduction about them, use the term more loosely. The fact is that to what extent clocks were made to stand on brackets which were specially produced with them is very uncertain, for naturally the bracket was the first part to disappear. There were, undoubtedly, some clocks made with their own brackets, but few indeed have survived that are not essentially table clocks for which there happens to be a bracket. It may be supposed that such a clock, so far from being a really luxurious piece for my lady's chamber, did dual duty as the bedroom clock by night and the house clock on the wall by day. There is also some tradition of spring-driven wall clocks which are fixed or hooked to the wall rather than free-standing on a bracket. One can only say that the 'bracket' clock does not derive in any simple or clear-cut manner from the weight-driven wall clock in which it might appear to have its origin. For almost as long as there have been domestic clocks, certainly from the fifteenth century, there have also been spring-driven domestic clocks—sometimes with horizontal dials like large watch movements on their backs, and sometimes clocks of the 'birdcage' and 'lantern' type. Moreover, the balance-wheel, which we might assume to be the hallmark of a portable clock, was originally used before the pendulum in wall clocks and was in fact virtually unused during the richest period of 'bracket clock' development in England.

From the time of the earliest long-case clocks—that is from about 1670—the portable clock, although its materials were similar and it largely reflected developments in weight-driven clocks, has had its own distinctive features. They usually originate in, if they do not continue to serve, its main purpose. Thus, from an early date, these clocks may have carrying handles on the top, though this feature is absent from those most directly associated with weight-driven patterns; you will not find it in lantern types or in some of the earliest wooden-cased bracket clocks which follow long-case hood design. The handles continue to be fitted throughout the eighteenth century, often on clocks of great weight and elaboration which one would scarcely dream of carrying by the handle even if they could be carried. Handles tend to disappear at the end of the century when, as with long-case clocks, new varieties of style appear, but they continue to be found, largely as ornament and often being rings mounted in the mouths of gilt lions' heads on the side of the case (as on the drawer handles of contemporary commodes and tables), until well into the nineteenth century. The frequent presence of alarms and repeating strike mechanisms seems evidence of the domestic function of bracket clocks and many were made which do not automatically strike the hour until a cord is pulled, when they strike the last preceding hour or quarter. Varieties of night clock also occur in which a candle light shines through a skeletal dial—but these are very rare and costly today.

Although the various striking and alarm systems have an obvious purpose, there seems no doubt that the bracket clock was very often an item of lavish splendour and expense, a status symbol maybe. Moreover, the limited space available offered a challenge to the craftsman. Hence spring-driven clocks from the earliest times have been designed to include all imaginable refinements and complications, particularly on the striking side. You will hear and see pictures of quarter repeaters and minute repeaters, systems which strike the hour a second time over in case you missed the first time, and alarm systems similarly arranged. You may encounter the logically ingenious which never really caught on—like the system of 'Roman' striking where, to save power, one note of the bell is used for the Roman V and another for the Roman I—and the elaborately useless for which there was some fashion—like the

Congreve clock of the early nineteenth century (where a steel ball rolls down an inclined plane and releases a catch so that the plane is see-sawed up into the reverse direction and the ball starts on its journey again), or the later 'beam-engine' clock of the Great Exhibition in 1851, where the escapement is linked to a visible beam swinging up and down in homage to the wonders of the steam engine. I say you will encounter them. More likely you will know them only by repute or in a museum showcase, for the extraordinary in clocks, whether useful and beautiful or not, nowadays changes hands at high prices and it is rare for a really antique curiosity to come your way at a price you are prepared to pay. Rather a couple of well-made and tasteful clocks of fairly typical appearance than a costly eccentricity will tend to be a preference forced upon you.

Granted, however, that the profusion of 'bracket clocks' is prodigious and the dating of them by external evidence (ie the maker's name) is likely to be more reliable than that of general appearance, there are still fairly well-marked features typical of spring-driven clocks in different periods, even within the category of 'bracket clocks'. Foremost of these is the top of the case (see Fig 16). Here the starting dates (though not the finishing dates) are quite well defined, though you always have to assess all the circumstances and, particularly, any signs of alterations, alongside the profile. For example, there are two well-known types of top, sometimes of lacquered brass, more specially gilt or silver gilt or even solid silver, known as the 'basket' and 'double basket'. The first is hardly to be found before 1680, or the second before the end of the seventeenth century. The 'bell-top' and 'inverted bell-top' do not occur before 1700, the arched dial not before about 1720.

The fashions in woods and dials are little different from those noted for long-case clocks. The round enamelled dial does not make a distinctive appearance before 1785, nor the completely silvered dial without separate chapter ring before,

Page 101 Mid-nineteenth-century French perpetual calendar clock. The calendar which takes account of leap years, is advanced every 24 hours by the striking mechanism. The case is not original

Page 102 (*left*) Early nineteenth-century French mantel clock; (*below*) Movement of French mantel clock, showing silk suspension

say, 1770. Because the pendulums of 'bracket clocks' do not beat seconds, or because the closest time-keeping was not their function (or for both reasons), seconds hands make a much later appearance than in long-case clocks. They are uncommon before the nineteenth century and never became the general rule. As with the long-case's lenticle, it was in the early stages

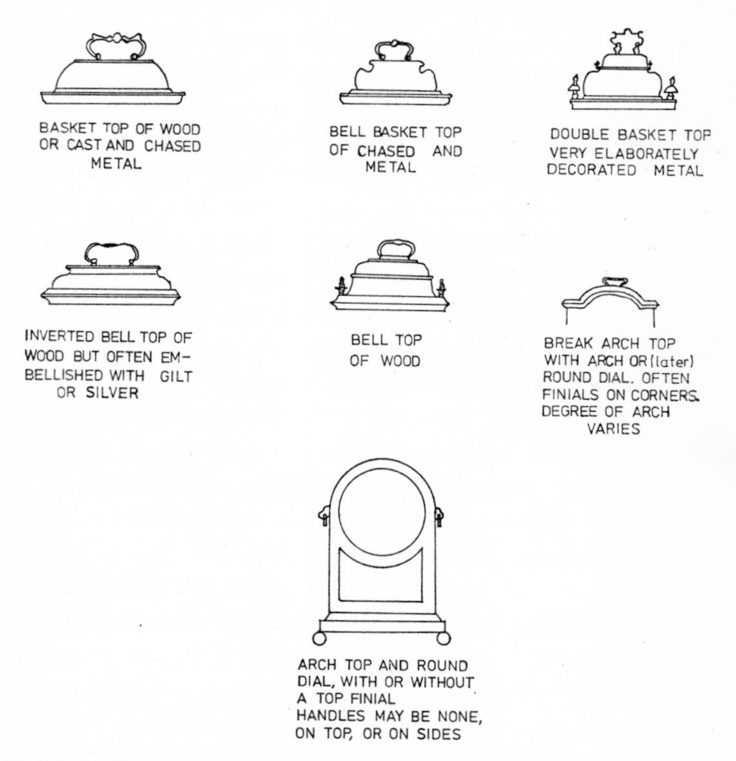

Fig 16 Outline development of bracket clock tops shown in chronological order from 1695. Note that many types overlap in date

desirable to have a visual indication that the clock was going, and many older clocks have a 'dummy pendulum', a small disc attached to the pallet arbor and showing through a curved slot in the dial above the centre. These are not a reliable indication of date because their decoration suggests an ornamental as well as a useful function. If, on the other hand, the slot in the dial is very much wider than it needs to be, this can be a superficial indication that the clock has been converted from a verge escapement or that the dial has been 'married'.

I think it would be fair to say that the average 'bracket' clock movement is better finished and with more attention to internal detail than is the average long-case movement which, even if it has the 'London' glass panels in the sides of the hood also found in 'bracket' clocks, is less open to display. This, at least of the early period, is another way of saying that the 'average' bracket clock is a great deal rarer than the average long-case clock. Not until the end of the eighteenth century does the greater plainness suggest that bracket clocks of a sort were to be found in a house in the normal course of events rather than as the cherished possession of the wealthy. Until the middle of the eighteenth century the back plates, bracket ('cock') supporting the crutch and pallets, and even the count-wheels of early clocks, were often elaborately engraved. Clocks from about 1730 onwards may have rear doors of glass; perhaps they were placed against mirrors to reveal their full splendour to covetous visitors. This fashion for engraving plates in an ornamental way died out slowly in the second half of the century, but many a lancet clock, for example, may have fairly rich engraving (these date from 1800–20 or so) and, even if the whole plate were not decorated, it was common to give the very edge of the backplate and cock a formal pattern.

The way in which 'bracket' clocks are fitted into their cases is a subject in itself and, of course, very much complicated by alterations and changes of case. The early movements tend to be screwed into the movement pillars from the base but, from the early mid-eighteenth century onwards, solid brass brackets were used, usually one on each side, one higher than the other, with large screws, which jut out conspicuously, through the case from outside. The 'bob' pendulum with the verge had a small wire hook on the backplate to which it was secured when the clock was moved. This bob was screwed up and down the pendulum wire to regulate the clock. The flat and heavier bob used with the recoil escapement was regulated either by a nut at the base, as in the standard long-case pendulum or, more often, by a short screwed rod attached to it and running parallel with the pendulum rod to which a lug with the regulating screw was fixed about a third of the way up. These pendulums have flat-sectioned rods and, for moving, were fixed into a clamp on the backplate. The securing hook for the 'bob' pendulum was to one side (usually to the left, seen from the back), and the clamp for the later pendulum is central on the backplate.

There is much more variety in size and elaboration here than in long-case clocks. It is generally true that the larger and more ornamental clocks are not the earliest. The taste for the monumental rather than the more purely functional grew, as in long-case clocks, during the first half of the eighteenth century. There is a line of development from here right until Edwardian times, if not later, when the vast case with its arched dial and impressive indications of strike/silent, fast/slow and, later, Westminster/Whittington, or eight bells/four bells chiming, is first and foremost out to impress and secondly to tell the time with a pleasant face and voice. As with the long-case, you can usually tell without much difficulty when clocks of this description were made. Often you will find inside all the walnut veneer (and maybe behind the name of some director of forty years' standing in a forgotten company), a rather pitiable little movement with stamped-out wheels and tuned rods for the chime. The face is constructed superficially in the traditional manner, but the spandrels are thin and flat, probably rough from the casting, and the chapter-ring figures fine and fussy, or maybe it has a silvered dial with no imposed ring. The minute and hour hands are virtually the same in pattern, and of thin steel usually painted black or blue rather than properly 'blued' (true blueing with heat gives a more translucent appearance than paint). Very often there is no carrying handle on top and at each corner there are turned wooden or brass knobs as finials.

You may, on the other hand, be pleasantly surprised by such clocks. You may look inside and find a pair of good thick brass plates and a coiled wire gong or two, or a set of bells and a musical-box barrel. Perhaps there will be a pair of fusees (for, though they have little purpose on the striking side, you do not find them singly) joined by a shining steel chain—the high-class equivalent of gut or wire. It may still be a relatively modern clock—the other signs will tell; but some of these so-called 'directors' clocks', presentation pieces, are very well made with solid movements, though they invariably need a good clean. You hardly want a house full of them, but one does no harm if you take a fancy to it. The same is true of good old cases with more recent movements of reasonable quality and for the 'English dial', where a large round enamel dial in a round case on the wall encloses a movement with pendulum and fusee. The English dial is too functional for many tastes, but the movements, which usually date from about 1875 to the First World War, though earlier ones, sometimes with verge escapements, are found, are generally excellent.

When you come to the smaller wooden mantel clocks you will have little difficulty in telling the old from the reproduction, but the exact period of a clock from the late eighteenth century onwards can be difficult to determine. The variety of types which sprang up at the turn of the century is enormous and many of the fashions lasted. With the help of a book on the subject, you will be able to date fairly closely a 'lancet clock' (which has a case shaped like a pointed church window in the Gothic style, usually with a round white enamel dial and the lions' head handles, and often superbly made and finished) or a case with a pyramid-shaped top chamfered out into little channels pointing towards a crowning brass pine-apple-styled finial (see Fig 17). Some research into the fashions

Fig 17 Later English bracket clocks

for inlaying with brass strip and the styles of fretwork and marquetry will be of assistance, as will a brief study of the history of hands. But when you are faced with a balloon case —a rounded top and tapered narrow waist—or the pleasant varieties of simple bow top with brass carrying handles, you may well be in difficulties. Since the details of all the wealth of nineteenth-century clocks are largely uncharted, and perhaps unchartable, you may as well make up your own mind when you have made what investigations you can. For, when all the investigation is over, with a 'bracket clock', however defined, it will be the qualities of the case and movement which are really important. A knowledge of these, let alone a good taste in them, can really only be acquired by experience.

FRENCH CLOCKS

So far we have considered only wood-cased clocks. But spring-driven clocks of course have cases of all sorts of other materials, particularly brass and marble. The commonest is the French clock, of which there are two main types readily available—the chased gilt clock, usually elaborate and under a glass dome, and the marble clock (the 'marble' may be imitation, alabaster or, if black, slate). These clocks have a bad name with non-specialist collectors, and for good reason, but my feeling is rather that they are God's gift to the genuine ham. The good reasons are two—namely that clocks of both sorts were turned out in hundreds of thousands towards the end of the last century and somewhat later and that, partly in consequence, it can be for the serious investor exceedingly difficult to tell the difference between a good one for his purposes and a bad one. The gilt style had, before its real mass-production, been made in various forms for over 150 years, and the 'marble' clock is not much younger. There is many an excellent forgery or good reproduction which is itself, by now, of some antiquity. In any case, the mass of French clocks imitate and extend their own traditions, so that even the detailed pattern of chasing or carving is not necessarily a reliable guide to age. There are books—too few and with insufficient detail—which survey the development and conventions of the French clock, and anyone who is primarily interested in the subject from the point of view of investment is extremely unwise if he does not search out what information is available.

These snares, however, are of peripheral importance to the ham, though naturally they affect him from time to time. The fact is that, at whatever date, there are good and bad French clocks around, and that for the vast proportion of them the definition 'good' and 'bad' is largely a matter of opinion and taste, particularly with regard to the case. There are, however, some guides to the rough age of these clocks insofar as their mechanism goes. The eighteenth century example, for instance, tends to have its pendulum suspended on a silk thread, which is wound up and down on a screw in the back-plate, instead of the more modern tendency to use a double slip of spring and to regulate by turning a small square in the dial—the silk suspension survived in French clocks long after it had been replaced by the spring in most other European clocks.

The great majority of older French striking clocks employ the countwheel system. There are characteristic differences in the styles of the hands and dials—the general rule being that the finer the figuring and hands, the older the clock. The winding square or squares which, in English clocks were laid out symmetrically level with the hands' arbor or below it, are from an early date in the older French clocks placed in positions which to English taste appear bizarre and eccentric. Most clocks which, by type, we should call 'French' were, at least as far as the movement goes, made in France, but many were imported and sold, and some were cased, in England. The presence of an English name on the dial does not indicate that the clock is not basically French, though it does usually indicate that it dates from about 1870 onwards. Most French clocks strike on small bells, and this is particularly true of the earlier ones. A coiled wire gong generally indicates a late clock. The same rough distinction applies to the case; those with bells tend to have a glass and bezel at the back as well as the front (for display purposes), and they are likely to be earlier than those with a pierced metal back-door covered with silk (to let the sound out and keep the dust away) which is a common arrangement for those of the end of the nineteenth century whether striking on gongs or bells.

Whatever these indications are for a particular clock, the chances are that it will be a good one for your purposes but you will have to be wary how much you pay for it. Most French clocks have circular plates for the movement and fine, friction-reducing pivots. The springs driving them are invariably in going barrels and are not powerful, for the wheels are small and well made, and the train will run smoothly with a minimum of power. The basic design of movement was unchanged for nearly two centuries. It required no modification because, like the substantial and totally different 'bracket' movement in England with its earlier start, it performed so exceedingly well. French clocks are almost always very well finished and beautifully made. The collector out for novelty and the investor with money burning a hole in his pocket naturally goes for the rarest and oldest—and often the distinction is only in the case. French clocks of many designs and varying dates are available both extremely cheaply and at staggering prices, and the rationale of the pricing is by no means always clear. If you can find one which you really like, you are almost bound to have a bargain. If you can find a well-proportioned clock with a visible escapement with dark red jewelled pallets, and with the perpetual calendar-work (taking account of leap years) patented by Brocot, you will have picked up a clock which is now worth money and will, before very long, be worth considerably more. If you fancy Mercury and Cupids and all the rest of the gallery standing on top of the case, or if you like your movement to be carried on the broad back of a marble elephant, you will have to pay for what seems to be a peculiarity of taste. But, whatever sort of case you fancy, do not despair if the clock whose appearance you really like seems to have a very conventional movement. These clocks are reliable, a pleasure to work on and their movements are not as standardised in detail as they appear from a first impression.

CARRIAGE CLOCKS

Also originating in France is the 'carriage clock'. The characteristic French carriage clock does not generally date from much before 1815, though there are analogous forms from which its development can be traced and genuine earlier examples can be found at a price or seen in museums. Its case is a brass (often gilt) frame enclosing panels of glass; the entire movement is visible. It is always spring-driven and its escapement, which may be lever, cylinder or a more exotic form, is mounted on top of the plates, across their edges, so as to be visible. As with French clocks in general, usually the earlier carriage clocks strike on bells, and later ones on wire gongs (though these are more widespread than they are in the larger French clock). The striking system is almost always the rack, but some of the earliest clocks use a countwheel. Most carriage clocks run for eight days, but many run for longer and the somewhat inconvenient period of a fortnight is not unusual.

There are many variations and complications in carriage clocks. A great many of them repeat the previous strike at the press of a button, and this may be extended to a system for repeating quarters and even minutes as well. The more excellent examples have jewelled bearings for most of the pivots. Carriage clocks with perpetual calendars exist, and some clocks were paired into a single case with a small barometer. The strike itself may be of the hour and half-hour which is usual in French clocks, or quarters on two bells or gongs, or may be 'grande sonnerie' (striking the previous hour each time before the current quarter). A large proportion of carriage clocks have an alarm, and this may ring on a gong or a bell (usually but not always the same on which the hours are struck). Most carriage clocks are rectangular, seen from the front, but they range from the enormous (ten inches high) to the diminutive (about one and a half inches high). Whether they are large or small, plain or chased, flat or pillared, they conform to the same styles very closely and this is one of the attractions of the miniature, especially if it is a striking clock.

Carriage clocks have, in the last ten years, increased very greatly in price, and it is now becoming difficult to secure a 'bargain', especially a broken bargain, in this line. The fact is, of course, that this type of clock suits everyone. It is reliable, a source of conversation and an ornament rolled into one, with the additional merit that its strike is too quiet to awaken you at night. It is, at the same time, unobtrusive to those who are not interested or impressed (for, alas, the carriage clock has become something of a status symbol in the living rooms of suburbia). It will not take you long to find the going price for the various types by looking round antique and jewellers' shops, but it may be as well to steer clear of central London and provincial city outlets unless you want to give money away. The advice can only be that if you see a French carriage clock going for below-average price you should not hesitate to buy it, because these clocks, unlike the ordinary French clock, are in demand and have a value which makes them well worth the attention of every professional repairer and restorer.

This last remark applies to all carriage clocks, regardless of quality and date. The odd situation exists at present whereby carriage clocks are priced rather vaguely in terms of how much work has been done to restore them and how complicated they are. Anything unusual—the miniature, the giant, the oval (which while being out of the ordinary run, is not rare), the calendar, the 'grande sonnerie', the seconds hand—all these departures from the average of the thousands of carriage clocks about command extra money, whether it comes from the ignorant who require a highly finished case and a beautiful strike, or from the collector who specialises in a particular oddity for its own sake.

At present simple age in a French carriage clock does not in itself necessarily command a premium. This is not because age is unrecognisable. Indeed, the older clocks have fairly distinctive features. For example, they have less brass and more glass, they have a larger base, chased or engraved, usually they strike on bells, they tend to have 'stop-work' to prevent overwinding, and they may have unusual escapements and a scapewheel pinion with twelve rather than the usual six or eight leaves. You will find few standard carriage clocks with solid brass doors, unless from a crude repair, but these are fairly reliable evidence of early clocks and are often finely chased with a simple pattern.

But these distinctions are not in themselves those which necessarily appeal to the large market for these clocks today, and there has moreover been little published on the detailed history of the carriage clock. Therefore it does happen that the humble ham may come across an old carriage clock very different from the run-of-the-mill productions from, say, 1880 till World War I, and have to pay no more for it than for a standard clock which happens to be in better repair. In due course it is likely that age will have more of a value and his clock will prove to be an investment. He has also minor advantages in that he may be less particular about the repaired result; a high-class jeweller or dealer will look askance at a hairline crack in the enamel dial of a carriage clock which cannot be satisfactorily repaired but which from the ham's point of view is scarcely a serious defect. Be mindful of this factor, however, for it can be a useful lever when the haggling starts for there is no really satisfactory repair. Sometimes brass sheet is used to mask damaged enamel round the edges, but new clocks were also fitted with such masks. There are now also gloss photographic reproductions for fixing over damaged dials if their appearance is intolerable. A chip or deep crack can be made less visible by filling with 'Porcelainit' or similar white enamel. But in the end it has to be faced that a damaged carriage clock dial cannot be made undamaged.

It should be said here that there are English carriage clocks. They are not the standard French carriage clock with an English name on the dial—for carriage clocks, like other French clocks, were often marked with the name of an assembler or importer, especially early in this century—but distinct pieces, usually with the names of well-known London firms upon them. They are almost always unusual or elaborate, their cases differ in style and proportions from those of French carriage clocks and many of them have silvered rather than enamel dials. Usually they employ a fusee and chain which is virtually unknown in the French version. These clocks command very high prices among collectors. They are rare and are unlikely to come your way in poor condition and at an acceptable price.

FOUR-GLASS CLOCKS

'Four-glass clocks' are like carriage clocks and of the same period, but larger as a rule (Fig 18). There is no clear dividing line between the two types of clock and a small high quality English four-glass clock might well, on occasion, be sold as a carriage clock. These clocks were made in England, with wooden cases and often striking on two bells, and in France from about the middle of the nineteenth century. The English versions have plain flat tops, glass or wood, the dial being rectangular, and the French for the most part have flat or rounded tops with a round dial behind a rectangular glass front panel.

Fig 18 English and French four-glass clocks

The English version, which may have a lever escapement or a pendulum and recoil escapement, is attractive and of good quality. It has an air of solidity, is plain, and displays the bevelled edges of very thick glass panels. The French version comes in all sizes, from some fifteen or more inches high to clocks no bigger than a large carriage clock. It is most familiar in a brass case, less square than English clocks of this type, with the escapement visible in front of the dial and often with a pendulum bob comprising two glass bulbs filled with mercury. This is an imitation of the true glass jar mercury compensating pendulum in fine long-case clocks and regulators; the mercury expands or contracts in proportion to variations in the length of the pendulum rod due to changes of temperature. The device in these French clocks is, however, more decorative than functional. These clocks are in fair demand, being reliable and pleasant to look at and the English examples, in particular, command a good price. A really dirty clock of these types looks an uninviting mess, however, and takes a deal of time to clean. Therefore there are still bargains to be had cheaply in the range.

FOUR-HUNDRED-DAY CLOCK

We are still in relatively unmapped territory and back with the contempt often given to the marble French clock when we think of '400-day clocks'. You will come across this type of clock quite frequently in junkshops and the like as well as new in the local jeweller's. It has a heavy rotating balance, usually, but not always, either a thick brass disc in the older clocks, or four linked brass balls in the newer versions. This balance is suspended on a fine steel wire and it is the torsion of this wire, according to its length and thickness and in conjunction with the diameter of the bob which can be adjusted by screws, which controls the time-keeping. To the top of the suspension wire is clamped a small brass finger which locates on a pin attached to the pallets. This is the equivalent of the crutch in a conventional pendulum clock. The escapement in all but the oldest 400-day clocks is a variety of dead-beat design very similar to that employed in Vienna regulators. The balance revolves very slowly, usually eight times a minute, and the mainspring is long and powerful with the result that these clocks, which were invented in 1880, achieve what was previously an expensive luxury—the need to wind them only once a year.

The great majority of 400-day clocks have a movement between rectangular brass plates (but sometimes round are used), mounted on pillars on a base-board, and under a glass dome. There is a device for fixing the pendulum during transit. Because the torsion suspension spring does not limit movement and is delicate with a relatively heavy bob—these clocks are even less portable than are most pendulum table clocks. Though this is the general pattern of the 400-day clock, there are other shapes and types of case and pendulum, both old and new. The hands are nearly always stampings, often ill-finished, and the dials normally painted or enamelled on the earlier clocks in the style of most French clocks of the same period. Later dials often have a silvered appearance (though they have not been silvered in the traditional manner), but the use of enamel continues to the present day. Some of the oldest pieces in this class have bobs of cylindrical and spherical shape.

Although they are collected, particularly in the United States, these clocks have a bad reputation among cognoscenti, and none of them can yet be classed as antiques. As a result they still tend to cost less when bought second-hand than do their rough equivalents when bought new. The reputation is something the ham need not necessarily support. They are, for the most part, quite reasonably made, though their finish is of little interest or subtlety. Their time-keeping is not irreproachable but, when they are properly adjusted, quite reasonable. After all, since they are not wound weekly, they are not set to time weekly either, and what appears to be a gross error may well have been accumulated over a month or two. Until recently there was an unavoidable inaccuracy owing to variation in the length of the suspension spring due to changing temperature, but springs are now available of a metal very little affected by these changes. I know that these clocks are entirely mass-produced and are not far off growing on trees. The fact remains that there is a considerable variety available, the mechanism and the smallness of the power supply intrigue, and it is up to you to pick and choose—one will cost very little more than another. You may well come across one with calendar work or strike, or with an unusual escapement or balance, and will learn a lot even from the more ordinary versions. They are not, at the moment, an investment proposition in this country, though you should be able to break even on resale. But the field of clock collecting is undergoing an enormous expansion as the older clocks come into the hands of those who want them for reasons other than having a serviceable clock about the house. Millions of 400-day clocks have been made, they are fragile, and one day the unusual ones at the very least will be valuable for novelty if not for craftsmanship.

SKELETON CLOCKS

The 'skeleton clock' is in much the same position nowadays. These clocks, with a structure of pierced and curved brass rather than of solid plates, and with a chapter ring only acting as a dial, were made in great quantities mainly in the second half of the last century. Some were hand-finished, and some were, in fact, hand-made apprentice-pieces. They vary from the simple timepiece, almost always with fusee and chain if British, through the striking clock of the simplest order which hits a bell once whatever the hour, to the three-train chiming clocks and musical skeletons, which are often of prodigious size and are sometimes known as 'cathedral skeletons' because of their Gothic cathedral pretensions. There are pieces with calendar work and novelties seeming to have only one wheel or no mainspring. They are always the talking-point they no doubt set out to be and in many instances command prices hardly related to their intrinsic value—that is to say, you could find a clock of similar complexity and possibly better finish or greater age, for a lower price. But of course intrinsic value is not easy to define where so much more than the functional is involved. You need to look around if you contemplate buying a skeleton clock. Ascertain the prices of similar clocks, or gain an estimation of the pricing policy of the dealer concerned, because many a dealer has sold a 'rare' skeleton clock to a tyro for an extravagant sum and there is very little system in the prices asked from clock to clock or place to place. But, if you are happy with the article and can accept the price, these clocks are not, as a rule, difficult to repair; a thorough clean will add pounds to the value and a great deal to your satisfaction, and even the simplest is no disgrace to any collection.

BUYING A SPRING-DRIVEN CLOCK

Finally, in a survey of spring-driven clocks which can do no more than help you to get your bearings, what is it prudent, when on the look-out, to avoid? The plain fact of the matter is that the clock which cannot be repaired or remade does not exist. It is a fact to bear in mind at more desperate moments, when a month's thinking about something else will often see you back with the solution to a small repair task which once seemed beyond you. The difficulty when you are buying clocks is to decide in the first place whether the work will be worthwhile and, in the second place, whether the clock will in the end retain enough relation to what it once was or what you would really like it to be.

There is, for example, no escapement which cannot be specially made but this work is not within most hams' scope and the people who will undertake it for you are few and far between. It is a long and very expensive job to make one escapement to size and the clock to which it is to be fitted must be of considerable value for it to be worthwhile. Meanwhile, if you have had a go at straightening out the existing escapement without success, perhaps the best course is to fit a

Page 119 (*right*) Early French carriage clock, c 1815; (*below*) Movement of French carriage clock, showing countwheel

Page 120 A group of French carriage clocks

good new mass-produced escapement and hope that in the future something approaching the original will turn up. The ham who has done all he can, or all he can afford, does not mind such an improvisation, because the one thing which he abhors above all others is the clock which simply does not *go*, however many pounds it may be theoretically worth. If the time comes when the clock changes hands and he has, after all, either to get a professional working on it or to drop his price, and he may be called an interfering botcher, he is nevertheless impervious to this, having already had his deepest pleasure.

The lever or cylinder escapement (called a 'platform escapement' in clocks because, at least until modern times, it is usually mounted as a unit on a small brass platform) is one thing and the pendulum escapement another. Spring-driven pendulum clock escapements do not differ in principle from long-case clock escapements, although some of them differ in detail. They cannot be bought as spare units but repairs are generally within your scope given time and patience.

The ornament on the older bracket clock cases and their spandrels can be a problem. Provided a sample remains, it can be moulded in plasticine or plaster of Paris, and then a replacement be cast in one of the modern hard-setting resin fillers. This can be painted with gold paint or restoration gold wax and will do until (if ever) the right piece can be found. The woodwork, as regards the structure of cases, is usually simple and presents no problem. Inlaid work can be given to a cabinet-maker depending on how large is the job and how valuable the case. Dials of 'traditional' construction can be restored without great difficulty but more than one attempt may be needed. The larger and duller enamel dials can be touched up with white enamel or filler tinted to match, but the damage will never be fully concealed. The delicate and glazed enamel dials (in fact enamel on a very thin sheet of copper) of many French clocks and carriage clocks have already been mentioned. They are very hard indeed to repair satisfactorily and the risk of making bad worse is always present. You can have a go, but the condition and importance of the dials of such clocks are factors to be considered carefully before taking on such a clock. With French clocks it not infrequently happens that the straps which secure the movement in the case (usually by clamping the case between the front and back bezels) are missing or broken, but their replacement causes no difficulty. The glasses of some of these clocks, though replacements are available, can be very hard to fit into their bezels. The safest course is to send the bexel to the materials dealer or repairer and have a new glass fitted rather than to risk buying a glass which is marginally the wrong size and which breaks when you try to snap it in.

As to movements, the difficulty of damaged wheels in weight-driven clocks has already been mentioned. The same tests apply to the running of the spring-driven trains, but damaged wheels are much more common in spring-driven clocks. You can replace one or two teeth with a fair chance of success but pinion leaves are another matter and so are missing wheels on the smaller clocks. Generally, in fact, size is a problem to take into account. To the clockmaker it is not so, because his skill adjusts to the situation, but to the ham repairs on spring-driven clocks are fraught with greater risks than those on long-case movements. The finer parts and better finishing (as well, in the older clocks, as the tendency to greater value) tend to reduce the level to which the ham should aspire by himself. Take a trivial but common example. The long-case pendulum hangs on a fairly substantial spring which, if it is broken or distorted, can often be straightened or shortened. The suspension spring of a French clock consists of two very thin springs joined by a yoke or mounted parallel in the same brass ends (Fig 19). These two springs must be of the same length and must not be kinked or distorted—as they almost always are when one obtains such a clock. The likelihood of making a successful repair to these springs is minimal, and distortion here can be the cause of bad time-keeping or even stopping, yet not be recognised. The springs are available cheaply, require very little filing to fit and the only sensible procedure is to buy a new one—as indeed does the professional. The same applies for the most part to a damaged hairspring on a balance-wheel.

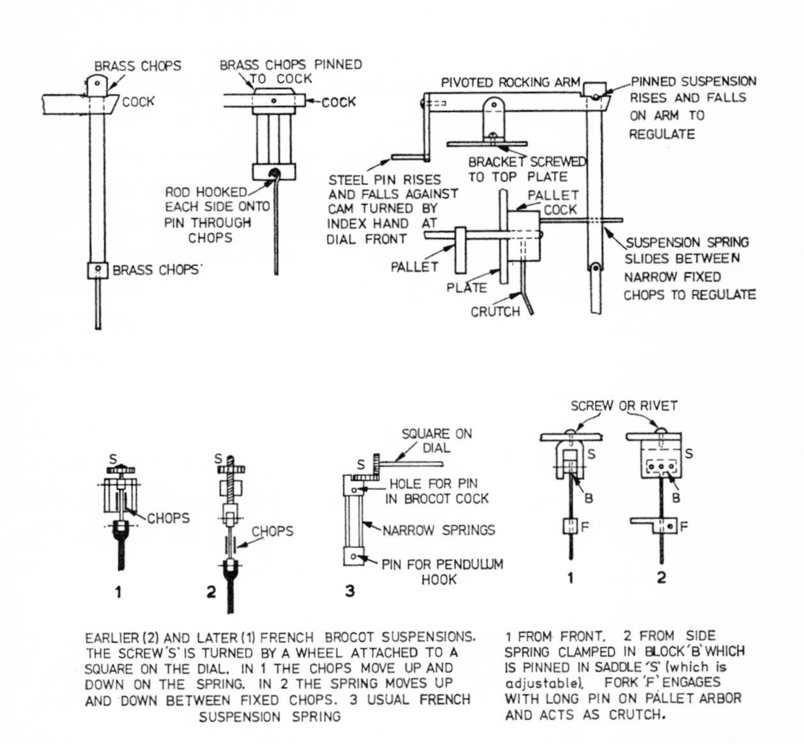

Fig 19 Some English suspensions (above), French brocot suspension (lower left) and 400-day clock suspension (lower right)

With mainsprings, the rule is that springs which are broken close to the outer end can usually be repaired. Springs which are broken at the inner end can sometimes be repaired but the job does not usually last and may harm the spring. Springs which are broken anywhere else must certainly be replaced. Replacement springs are available, but it is wise to send the barrel with the order to ensure that the correct size is fitted. A going barrel with missing or damaged teeth is a serious risk to take on. One tooth can often be repaired or replaced, but it is a chancy operation when you bear in mind the strains to which these teeth are subjected. Normally you cannot undertake to cut a fusee or missing striking rack; whether it is worthwhile to have these jobs done depends very much on the clock concerned.

Missing parts in platform escapements cannot, as a rule, be replaced from stock, but they can be specially made at a price. Normally, however, a replacement platform is the best answer. French clock pendulums present few problems. You can buy them to size new, or you can pick up another French clock at junk price, keep the parts, but use the pendulum, adjusting it or the suspension as necessary for length. Many French clock pendulums are more or less interchangeable.

I have outlined the difficulty of carriage clock dials. Carriage clock hands are available new, though you will have to rivet on a brass boss ('collet') if one is needed, and you may have to settle for a compromise in the style of hands. The older types, pierced at the tips, are hard to obtain. Many a carriage clock will come to you with missing or chipped glass panels. There is nothing more unsightly. You can buy carriage clock panels from the material dealer if they are plain glass, but it is vital either to be exact in the measurements quoted or to send the case for fitting. The valuable painted enamel panels of some older clocks cannot be replaced.

The torsion springs of 400-day clocks (Fig 19) can be bought, but they are expensive. If you are likely to have more than one of these clocks to deal with it is worth buying an assortment of springs and there may be no alternative. If you are likely to require only one such spring, consult *The Horolovar*

400-Day Clock Repair Guide (published in the United States by the Horolovar Co). You will very likely be able to identify your clock by the pendulum bob and the design of the back-plate illustrated there and thus ascertain the required suspension. You can also find out by experiment with a set of springs, but this takes time. For the standard sizes of these clocks glass domes can be bought new. For all others, and also for skeleton clocks which often have specially shaped glasses, you are really dependent on a constant vigil, especially where stuffed birds are known to nest from time to time.

When all is said and done, the ham who buys an old clock takes a considerable risk. Oft-times, though he knows very well that he is a fool to do so, he is so blinded by his vision of what might be that he takes a clock off a dealer or advertiser, or as the sole item he requires in a junk lot at an auction, when he has had no chance to examine it. Then, even with the keenest eyes in the business, he cannot spot everything. Clocks, like cars, are sold for other reasons than that the owner wants a change or has died or needs the cash. There can be few of us who have not at one time or another taken in a clock with several missing pinion leaves, made a totally incorrect assessment of the style and date of a case, or discovered too late that case and movement have not been together all their lives. The results of these errors tend to be two-fold; you stand to lose money, and you stand to lose the pleasure of doing the job yourself.

It is obvious that one cannot specify how these calamities are to be avoided because so many variable factors are involved—your own experience and equipment, how much money you have spent, how much you should have spent, how good is your source of materials, and so on. But it will be clear, too, I think, that very few such blunders amount in the long run to absolute disaster. You may, in time, be able to do what you thought you could not do; you may be able to find someone who will do it for you, or, at the worst, you will accumulate the remains for possible future use. I have tried to indicate the areas in which one does, in general, need to tread a little warily. I want, in the remaining chapters, to indicate some of the repairs which you can undertake with considerable hope of success.

6

Washing up and stripping down

CLEANING

It is often said that there are two main reasons for cleaning a clock—to make it look nice and to help it to go properly. Dirt is ugly—you will doubtless have seen carriage clocks done up for sale and carriage clocks which seem never to have been cleaned in their lives and have noted the difference—and dirt breeds wear and friction, which work slowly and remorselessly on a clock rather like woodworm or dry rot in the timbers of a house. These two reasons for cleaning are sound ones, but I would add another at least as important. It is only when you strip a clock right down and contemplate it with the slowness essential to proper cleaning that you really get to know the clock. You may discover something important, like the name of the maker on the back of the dial or unnoticed holes indicating an alteration or a conversion. You may also discover the skilled and unskilled trademarks of predecessors in the form of good or abominable repairs. But, above all, you are bound to discover how this particular clock works or should work for, though the principles commonly met with are few, their detailed applications vary enormously.

Personally, I make it a rule that I never take over a clock and, indeed, rarely work on one for someone else without giving it a complete clean. If it is a long-case clock movement, or that of a French clock almost totally sealed into a tight-fitting case, it is true that the works are not normally visible. But you will not be able to spend many months without a look at the heart of the matter, and a great deal of your pleasure will be lost if you are met by a mass of tarnish, grease, or worse. Moreover, half the parts will not be discernible at all clearly and you will have no sure knowledge that they are working exactly right. It does not follow that everything will be highly polished and shiny—many of the large movements housed in cases without glass panels were never polished or highly finished except on the working surfaces where friction has to be kept down. But there is a world of difference between clean unpolished brass or steel, and blackened or even rusted metal. You will want to know your clocks inside out and want them, as far as possible, in the pink of condition and seen to be so. The only way is to clean them.

Cleaning a clock is a slow job and it will not be hurried. You will not produce the finish you want, nor even improved mechanical efficiency, by plunging the movement into a bath of petrol, letting it dry off, and then applying oil liberally to keep things moving. If you give this treatment to a mainspring barrel you will do actual harm because the spring will not dry out properly. There is in fact no satisfactory alternative to taking the movement completely to pieces and cleaning each part one by one. Beware that it takes time and plan accordingly. Unless you have the space to leave parts lying safely around for days on end, you need a long clear period in which to get down to it. That way, you will not have the frustration of picking up unrecognisable pieces from the floor because the children have kicked the table, or the problem of keeping dust and dirt-free the parts you have already cleaned. I make cleaning a weekend job. I find a timepiece can be covered fairly well in a day and a striking or chiming clock can just be managed in a clear weekend.

We are led to believe that matter, like energy, cannot be destroyed. Anyone who has cleaned an untended old clock will bear witness to the fact. Dirt is not destroyed; it is merely moved from one place to another. In the interests of domestic harmony and to avoid undoing all the good work, you must provide a home for what comes off the clock. You need an overall, a vast quantity of rags, two or three jamjars and a large receptacle for bits to soak in—the sort of tray made for cars to drip oil into does well and so does a big baking tin. The essential cleaning materials are household ammonia and soft soap or washing-up liquid and metal polish. Window-cleaning liquid or spray is a help on stained glasses. By way of tools you need your finer files, emery sticks or paper, buffs or wash-leather, wire brushes, pliers, and an assortment of screw-drivers.

There is no mystery and little skill in the cleaning process, although everyone naturally has his favourite method and practice varies according to the quality of the work and the nature of the dirt. Cover the bench or table with white paper if possible; it is only too easy to lose track of screws and small parts and the paper can also be used to sketch the positions of pieces as they are removed. Pour ammonia into the tray and a jar or two and dilute it to make it go further—how much you can dilute it depends on how resistant the dirt and tarnish are. The dilution should be with very hot water because the solution works best when warm. To the diluted ammonia add soap or two or three good squirts of washing-up liquid. During this setting-up it is advisable to wear rubber gloves because ammonia is painful in any open wound, and to work with windows or door open because the fumes are unpleasant and cause the eyes to water. Instead of the ammonia mixture you can, of course, use commercial fluids, but they have as a rule no advantage for cleaning by hand and their constituents are in any case much the same. You can also use paraffin or petrol, which are effective in removing grease and old oil, but they will not of themselves make much impact on tarnished brass.

Into the solution you immerse all the brass parts (and this includes those, like wheels, with steel fittings), but do not give the treatment to the spring barrel, if there is one, until you have removed the spring. Small parts can be strung on a piece of wire in one of the jamjars. Be very careful that everything you put into the solution is completely covered; if it is not, there will be a nasty highwater mark which will be very difficult to remove. How long you leave things cooking is up to you. The job will usually be done in ten minutes or so, though large and very dirty parts require longer but you will no doubt be getting on with some other stage of cleaning meanwhile and the process does not need to be timed with any precision. Do not, however, leave them in for much over an hour, or there may be stains, especially if the metal is not of even quality. Personally, I clean as I dismantle, cleaning up screws, hands, and other small pieces while two or three brass parts and wheels are soaking. Then I put another batch in while the first parts receive the next treatment. The plates I put in towards the end, but not right at the end, of the ritual, because they tend to need a longish soak and are always the job which looks quickest and turns out to be longest. Incidentally, if any of the parts which you soak in this fluid are lacquered or varnished, the lacquer will come off, so you need to decide whether to leave well alone, whether to relacquer afterwards, or whether in future these parts are not going to be lacquered at all.

Basically, the ammonia solution loosens dirt rather than removes it, though at the end of the day there will be a nasty sludge at the bottom of jars and tray and the fluid cannot profitably be used more than once. In fact, when you remove the parts after their soaking, you may be alarmed by their appearance—often they turn a dull grey or brown which looks a good deal worse than before they went in. All is well, however. Rinse off the solution by brushing with a brush soaked in benzine or petrol, since ammonia remaining on steel will cause corrosion. You now need to dry the parts off, either in a very gentle heat, such as from a fan heater, or in a lint-free rag. If there is still a lot of dirt about, and if they are not highly polished parts, give them a good hard brushing with the brass wire brush; in any case do this to the teeth of gear-wheels (it is here that the power unit and rotary brush mentioned in the second chapter come into their own). Highly finished pieces, especially those from French and carriage clocks, are liable to be scratched with the wire brush on their smooth surfaces and with them it is best to go straight on to the final stage of rubbing with metal polish. Once this has dried, you can buff the parts with chamois leather. Tiny nooks and crannies can be cleaned out with sharpened pegwood. The wheel teeth will need a final clean with a soft brush. When these brass parts are cleaned avoid, as far as possible, touching them with the hands; what at first looks like a dull smear from the grease in the skin in due course turns into an unsightly brown stain, and it is always much easier to clean a whole part (particularly a plate) than it is to touch up a spot on it at a later date. Cover your fingers for the larger parts and use tweezers for the smaller ones. The professional handles polished parts in tissue paper and avoids the stain problem.

The treatment of steel depends on the state it is in. Well-polished pieces will usually do with a good rub from a rag with a little oil on it. Slightly rusted parts respond well to the wire brush. The more severely rusted pieces may need work with an emery stick or even a fine file. Only in the worst cases will you need to resort to rust-removers or acid; but then do take precautions like wearing rubber gloves and keeping the container in a safe place and closed when not in actual use. Parts which receive this drastic treatment will be clean and reasonable in appearance. They will not spread rust in future provided the cleaner has been thoroughly washed off and they have been fully dried, particularly in the crevices. According to the severity of the rust, however, they will be pitted in places. You cannot deal with this in round parts unless you have a lathe, or a motor with a chuck attachment, in which case rotating the parts against files and emery sticks of increasing fineness will eventually reveal new and true metal. Flat surfaces can be filed and buffed clean and smooth.

Be particularly careful in attacking pivots and steel hands. Hands will probably have to be reblued which demands a good clean surface. Pivots have to fit their existing holes or have the holes altered to fit. The really rusted pivot has to be renewed rather than cleaned but, for the rest, the vast majority come out of the wash, after a rub and brush-up, as clean as you are likely to get them. To tamper much with pivots, particularly fine ones, is rather to invite trouble, at least for the beginner. However, where a pivot is not seriously out of shape but only somewhat jaded, it can be cleaned by rubbing with a burnisher or the flat side of a file which has been slightly roughened with fine emery paper. This job is most easily done with the arbor held in a chuck and the pivot supported on a wooden block with a little notch for it to rest in.

Pinion leaves can be cleaned out with the folded edge of emery paper, with or without a pointed piece of pegwood inside to stiffen the edge. Mainsprings, like fine pivots, are best just wiped clean and then wiped again with an oily rag held between the jaws of fine pliers. Piled-up oil can be removed with petrol or paraffin. On no account uncoil a spring to clean it; the inevitable distortion will spoil its action and may lead to a breakage in future. Pendulum suspension springs should be wiped clean. It is no use filing or scraping them, because the slightest distortion here will ruin the timekeeping if it does not stop the clock. Suspension springs are cheap and the cure for a rough one is replacement. The same applies to hairsprings, though dirt, as opposed to rust, can be removed by dabbing with fluid, rinsing, and then folding in pieces of paper tissue to dry, followed by a very light brush.

A movement done up with new pins looks a great deal better than one into which old pins, imperfectly straightened and with gouges out of them, have been fitted. Such small touches cost little and make all the difference. Filing off burrs on winding squares is another detail and attending to screws is yet one more. Old screws of a distinctive shape, especially square-headed ones, can be cleaned up with the wire brush and emery, or the rust-remover treatment. A more modern screwhead, which may be polished, and which may have been badly burred by careless use of a screwdriver or by forcible extraction when rust has jammed it can be difficult indeed to restore. At the outset, naturally, you do your best. Later on, you will probably have built up a stock of screws and will find one the right size and style. There are, however, desperate situations where one fits a set of new screws simply because the missing one cannot be matched. There are also occasions when a problem can be solved by enlarging the screw hole and tapping it for a larger screw from the parts box.

The cleaning of clock plates always takes longer than expected, but it is also rewarding in their final appearance. When they come out of the solution they should be dried and then treated with metal polish. It is essential to clean out all the holes, and this is done by twirling in them a sharpened stick of pegwood, off which pieces are successively shaved until it eventually comes out of the hole perfectly clean. Oil sinks for retaining oil at the pivot ends (shallow cups surrounding the holes in good-grade movements) can be cleaned with thicker, rounded pieces of wood, or with pegwood capped with a piece of cloth. It is also possible to buy highly effective shaped buffs for use in the chuck for the purpose; without a power unit you can make good use of such a gadget or the pegwood held in the chuck of a hand-drill. Engraving can be cleaned with pegwood. If you want a really clean plate there is no choice but to go over it at least twice; at whatever stage you buff off the dried metal polish some will find its way into the holes, and, if you leave the holes till last, some of the congealed oil from them invariably smears the surrounding plate. A useful and time-honoured arrangement for cleaning the awkwardly-shaped holes in the plate, such as the one the crutch goes through (and also for cleaning the crossed-out sections and inside rims of wheels) is to tie strips of chamois leather and rag to the table or bench by one end. A strip is then pushed through the hole and the part is rubbed up and down, and the strip held taut until it is bright and clean. Unpolished plates often have, or can be given, a pleasing grain. Get them thoroughly clean and then place them against a firm edge or bench-hook. Rub them slowly and hard, dead straight and in one direction only, with decreasing grades of emery paper held round a block of cork or wood. The holes will of course need cleaning afterwards.

When it comes to the dial, tentative experiment is best. The problem is not only to clean the metal but also to avoid removing the figure-work. Immersion in cleaning solution is almost always detrimental. Go easy with abrasive and brushes! Chapter rings are best washed in soap and warm water; metal polish does not always clean them and, when it does, it removes a fine layer of silver, so that it is not advisable to polish hard a thinly silvered or patchy chapter ring. Damaged silvering can, however, be reasonably restored and the process will be outlined later. White enamel dials are not usually difficult, because their glazed surfaces only permit dirt to adhere rather than to penetrate. A wash-over with a little milk, with bread-crumbs, or a light polishing with window-cleaning liquid will usually do the trick. It is also possible to buy from the materials shop a special rubber for use on them. But be careful with these dials, the figures are not recessed and will scratch off if provoked, and the enamel is extremely brittle—a slight bend will seriously crack or chip the dial.

So much for the cleaning process. In outline it is common to all clocks and presents no major difficulties. It is slow work, but the reward of a shining mechanism running smoothly and precisely is ample. There are of course minor snags and dangers. Some have been mentioned and you will soon find others. But this is a field where invention has full play and you may roam profitably round the kitchen cupboard or chat with the chemist's dispenser, discover his personal answer, and bask in the knowledge that you have discovered something. Years later, you will very likely find that the obvious way was the best, or that the discovery was known to the makers of old and is listed in the browned pages of the many nineteenth-century manuals on clock work; but no matter, if it has served your purpose.

STRIPPING GENERALLY

Before cleaning, however, comes the stripping, which is not common to all clocks and has the much greater danger that if it is not carried out properly there will be additional large repairs to be done or a multitude of parts scattered about the room which you will despair of ever reassembling. When you strip a clock down for cleaning and repair, do the job slowly and with care. Your aim is to see how matters should be, how they are, and how best to avoid making them worse.

Removing power

The first thing to do with any clock ready for stripping is to take off the power. A spring-driven clock fully wound and liable to fall apart is a danger to itself and to others, a weight-driven clock complete with weights and winding gear is merely inconvenient. Take first the weight-driven clock. Suppose that you have made the preliminary inspection outlined in the third chapter and you now have a long-case clock, or perhaps a Vienna regulator, ready for attention. When you take anything heavy, be it pendulum or weights, off long-case movements, you upset their balance and they are liable to topple forward because, though these movements are usually screwed to a 'seat board' in the case, the board itself usually just rests on the sides of the case (the 'cheeks') so that the exact position of the movement and dial relative to the hood can be adjusted. The proper procedure is to remove the hood of the case, to place one hand on top of the movement and with the other to remove the weights and the pendulum, easing the suspension spring gently out of its cock and drawing the pendulum out carefully downwards through the crutch. At this stage, check the state of the suspension spring and of the drive wires or guts; replacements for these items may as well be in the post while you are attending to the rest.

The Vienna regulator and similar types of clock do not have a removable hood and their movements are fixed firmly to the case, usually being clamped to iron brackets protruding from the backboard. You do not, therefore, need to support the movement but you do need to be especially careful with the pendulum because its suspension is more delicate and is unlikely to be visible. It may hang from a cock in the usual way on the movement, but more often the cock is part of the casting on the backboard. In this instance, once you have removed the weights, unscrew and slide out the movement before you tackle the pendulum.

Removing hands and dial

Your next step is to remove the hands and dial. The minute hand will be secured with a pin and washer. The hour hand may be a press fit on its arbor or it may, in addition, have a small screw. The dial will be distanced from the front plate by pillars ('feet'), usually three or four, passing through the front plate and secured by pins. Be careful in extracting these (and all pins)—not for the sake of the pins, which you will replace, but because a slip here can leave a nasty scratch on metal which will in due course be polished. If, however, a pin is broken off short, there may be no choice but to knock it out backwards with a punch as best you can. Many Vienna regulators and continental clocks of the last century have complications in the dial fixtures—one or two sub-plates are secured by pins, and a centre held in position by clamping screws, for example. These, though they are a nuisance if you need to see behind the dial in a hurry and are not always easy to reassemble, do not present any great problems at this stage. Neither does the simple fixing of the dial in 'birdcage' movements—here the dial often has lugs fixed to it and these are merely screwed to top and bottom plates.

Striking work

When you have the dial off any clock you will be confronted with a mass of mechanism which you had perhaps never realised existed, for often it is completely concealed by the dial and the edge of the case. The main components are ancillary wheels, which reduce the speed of the minute hand's arbor to that required for the hour hand and actuate any alarm and the strike, and included here is usually most of any calendar work and suchlike that there may be. Here also is the rack, if that is the kind of striking system. The exact arrangements of the rack mechanism differ slightly from clock to clock, but you will always find the rack itself, the snail (on the hourwheel or mounted separately) onto which it falls, the rack-hook which allows it to fall or prevents it from doing so, and the lifting piece, lifted by a pin or cam on one of the ancillary 'motion' wheels which sets the whole mechanism off (see Fig 20). If the clock is a single-handed one, there is usually only one wheel behind the dial, and it usually engages with a pinion on the arbor of the going-train pulley. Here the striking mechanism is set off by a twelve-pointed star fixed to or driven from the hour wheel, and the lifting piece is so shaped that it is pushed up by the teeth of the star. The countwheel system employed here has of course no rack or detent—only the lifting piece is at the front of the clock.

Whatever the striking system, and whether the clock is spring- or weight-driven, this is the last opportunity to see how it works or should work and you will do well to make the most of it. Temporarily replace the minute hand on its arbor, or turn the arbor with a suitable key and watch how the lifting piece is raised by the motion wheel. Keep a hand on one of the early wheels in the striking train again so that there is power there, and you will see that, just before the lifting piece has risen as far as it can, the striking train will run momentarily and then come to a stop. At the same time, in the rack system, the rack will have been released to fall onto the snail. Now look between the plates at the striking train and you will find that a pin on one of the wheels (the 'warning wheel') has collided with the other end of the lifting piece. Once you have turned the hand on past the hour, the train will be free to operate. It is this 'warning', and the rack falling, which you hear three or four minutes before a clock strikes. Keep the pressure on the wheels and a pawl, known as the 'gathering pallet' and attached to the extended arbor of another wheel in the train, will turn and, tooth by tooth, lift the rack up as the hammer blows are struck until it is once more held (in old English clocks) by the locking of this pallet against a steel pin on the end of the rack. Alternatively, there will be a pin on the wheel on whose arbor the gathering pallet rides (the 'locking wheel') and this is in the path of a locking piece on the same arbor as the rackhook; the locking piece will drop back into its normal position once the rackhook has fallen into place to hold up the raised rack, and then the locking piece will collide with the pin on the locking wheel and stop the train. This is the usual arrangement in French clocks. There are differences in detail according to the date and origin of the clock with rack striking, and a special system is employed in many carriage clocks, but the general principles are common to all.

In the clock with a countwheel you will find a similar lifting piece and 'warning' system, although, again, the details vary (see Fig 21). The number of blows struck is determined by the falling of an arm into slots on the countwheel. This must, however, coincide with the falling of another finger on the arm into a notch on a wheel in the striking train which revolves once for each blow struck. In English clocks this wheel, which is the locking wheel (for the countwheel does no locking itself), has a raised edge of brass riveted round it, with a gap at one point (hence its other name of 'hoop wheel'). When the counting arm falls into this notch the train is held up. It will always fall back into the notch unless it is prevented from doing so by the raised sections of the countwheel on which the other end of the counting arm is operating. As with the rack system, when the lifting piece is raised by the motion wheel, the train is momentarily freed from its locking but is immediately held up by the collision of the warning-wheel pin and the lifting piece, and it will only run freely and strike the hour when the lifting piece has fallen, that is, exactly when the hour is due. The French countwheel system is similar, except that the locking wheel is not a 'hoop' wheel but akin to that in the French rack strike locking arrangements. French clocks usually strike one blow at the half hour; this is arranged by having an extra lifting pin or cam, in the motion work. In rack striking clocks, this pin is so placed that it does not fully raise the lifting piece and the rack does not fall. In the countwheel system, the countwheel has extra-wide slots between the hour sections

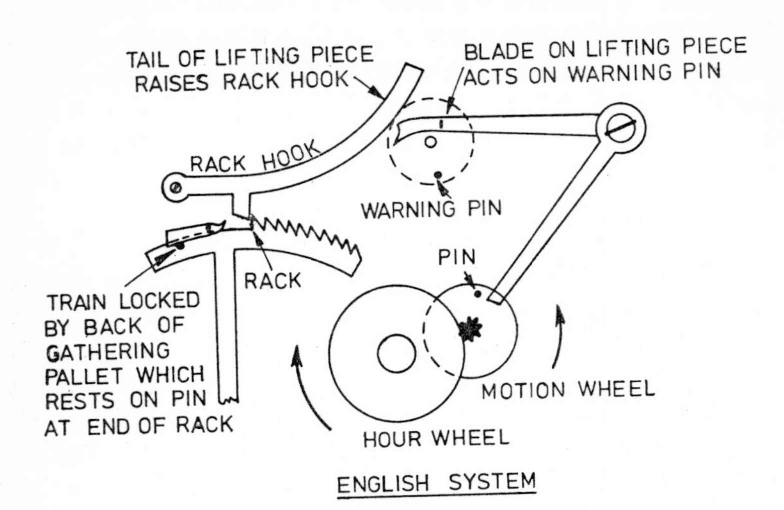

ENGLISH SYSTEM

GATHERING PALLET

WARNING PIN

HOUR WHEEL

CANNON PINION

PIN ON RACK HOOK RAISED BY LIFTING PIECE TO RE-LEASE RACK. RACK HOOK, ON ARBOR OF LOCKING PIECE, RELEASES TRAIN WHEN RAISED

FRENCH SYSTEM

Fig 20 English and French rack strike release systems

and there is no raised section after the half-hour space to prevent the strike locking after one blow has been struck.

If the clock is a normal quarter-chiming movement it will

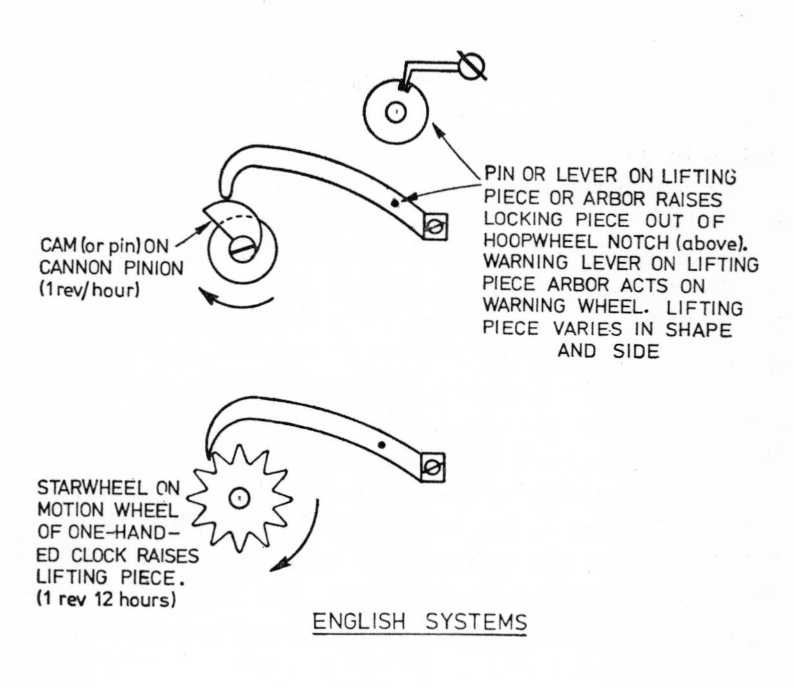

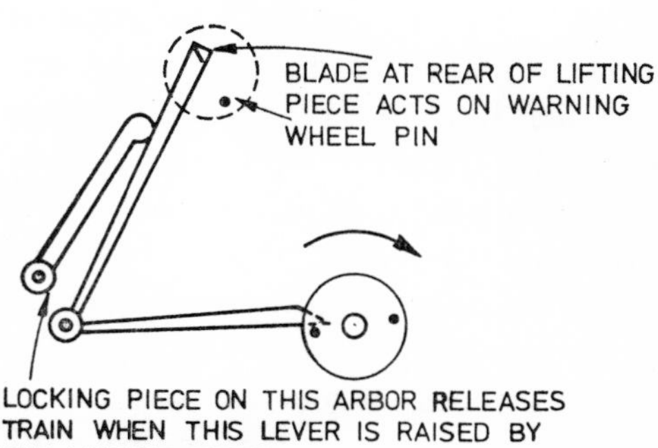

Fig 21 English and French countwheel strike release systems

have a third gear train and the whole rack system will exist again virtually in duplicate, the second rack having only quarter divisions, and falling onto another snail, divided into four. When the chime has done its piece it finally sets off the strike mechanism at the hour, this having been held waiting at the warning position during the quarter chimes. Alternatively, in the more modern arrangement, the chiming is controlled by a countwheel on the front plate; on this is a knob or pin which releases the striking when the fourth quarter is struck.

French carriage clocks striking quarters, and other clocks striking 'ting-tang' quarters on two or more bells or gongs, however, have only one striking train. Usually there is a quarter snail and a separate rack mounted loosely on the same arbor as the hour rack. The two hammers merely operate off the same pin-wheel. There is a device for silencing the hour strike until the third quarter has been struck, for preventing the striking of quarters and hours together, and for silencing one of the hammers and gongs while the hour is being struck. As the hands are moved round you will see that a lever is raised to prevent the hour rack from falling until it is wanted, and another small lever whose tip prevents one hammer from striking at the hour (see Fig 38). In the 'grande sonnerie' there is another lever, manually operated, for preventing this device's operation and in consequence these clocks, which have extra large mainsprings in the striking trains, can strike the hour before each quarter if required. There are some ting-tangs which work by a different principle; here there is one very elaborate snail, with a step on it for each quarter as well as for each hour, and only one rack is used, its first three teeth being short. Again, you will find movements with two snails and one rack. Provided the main principles of the French version are understood these variations offer no great problems.

Carriage clocks and many English striking clocks mount

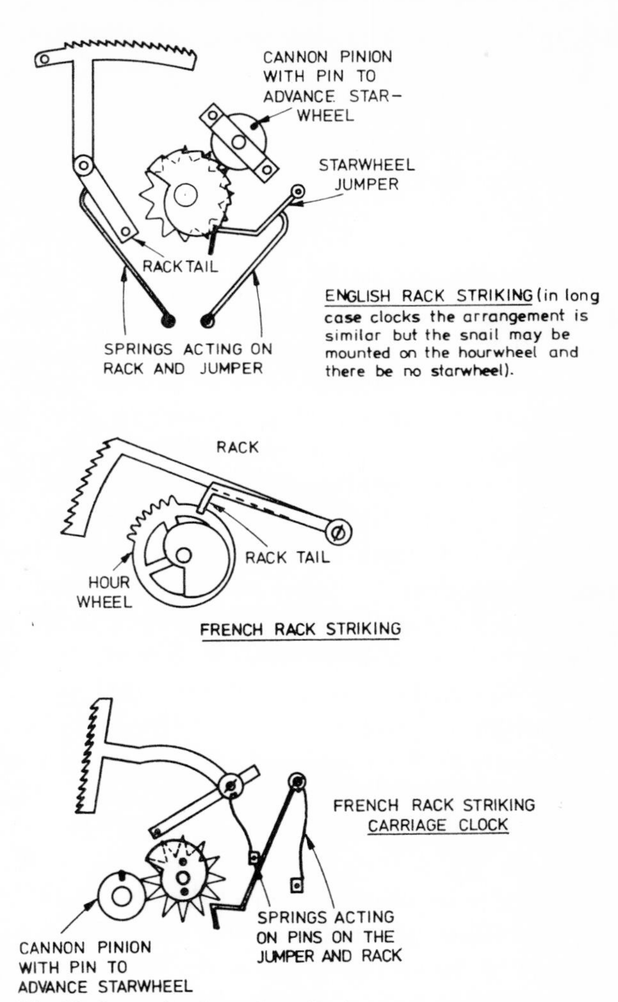

Fig 22 Star-wheels and snails in rack striking systems

the snail on a pin fixed to the plate and a star-wheel, with a sprung arm (or 'jumper') to hold it in position, is fixed to the snail (see Fig 22). The whole assembly is turned by a pin on the minute wheel which catches on a tooth of the star-wheel. By this means the snail controlling the number of blows struck is advanced not continuously but by a series of jumps immediately before the hours; thus the repeating clock, if correctly set up, will always strike the hour previous when the repeater mechanism is worked because the snail will not yet have been moved forward.

Motion work

The remaining matter on the front plate of your clock will be the motion work, the means of reducing the speed of the centre wheel and minute hand so that the hours can be shown (if it is a two-handed clock) and of making sure that you can set the hands to time. If nothing were built in to cope with this latter point, you would have to take the hand off and move it round, or else wait for the right time to come round for you to start the clock again; trying to push back a fixed hand would break the hand, if you were lucky, or the wheels and escapement if you were not. Friction, so often to be avoided, is here exploited. The hand of a striking clock must be fixed to its own wheel because that wheel is directly connected to the striking system which must always be in the same position relative to the hand, but the wheel and hand can be adjusted in position relative to all the other wheels in the clock. The particular arrangements to this end vary (see Fig 23). One of the commonest, and almost universal in long-case clocks, is for the minute hand to fit onto the square end of a brass bush and wheel (known as the 'cannon pinion'). The cannon pinion may be friction tight on its arbor (which is that of the centre wheel of the clock) or it may be loose, but tensioned by a nearly flat spring plate behind it; in this latter case you will find the hand and pinion have to be pushed hard back against the spring until the securing pin can be inserted through the hole in the arbor in front of the hand. The friction or tension are important. They must not be so great that the safety of the hand is at risk when the clock is set to time but, if they are too slack, the minute hand may fail to go round regularly and the striking action will be upset also. In more modern clocks, and often in the later carriage clocks, a rather different system is employed. Here the actual central arbor is in two sections, held by a spring against each other or, alternatively, the centre wheel of the clock is not fixed to the arbor but held tightly tensioned onto it by a spring. Whatever the arrangements you will have no difficulty in understanding them provided you know that the purpose is two-fold—to turn the minute hand in the normal course of running, but to allow it to be moved independently of the main train when setting to time.

The hour hand and its wheel are mounted on a pipe over the cannon pinion and, whether or not they have a supporting bridge or bracket, they must be quite free to move, subject only to the motion work gearing. Where there is only one hand and motion wheel the arbor may be pivoted at one end into the front of the front plate and held in position against the dial by a tensioning spring which permits the arbor to be moved independently of the wheel, or the wheel itself may be secured to the arbor by a spring clip.

Pendulum and escapement

At the top of the movement, between the plates of a pendulum clock, you will see the escapement wheel and the pallets, attached to the crutch, mounted above it. One end of the pallet arbor is usually pivoted into the detachable suspension cock for the pendulum, if that is hung on the movement rather than on the backboard. This cock may have elongated screwholes so that the height of the pallets above the wheel (the 'depth' of the escapement) can be adjusted for proper working. Alternatively, it may have been exactly positioned with steady pins as well as screws to locate it, in which case the adjustment, if needed, is less straight-forward. In most such movements the bell is mounted on a stout bracket

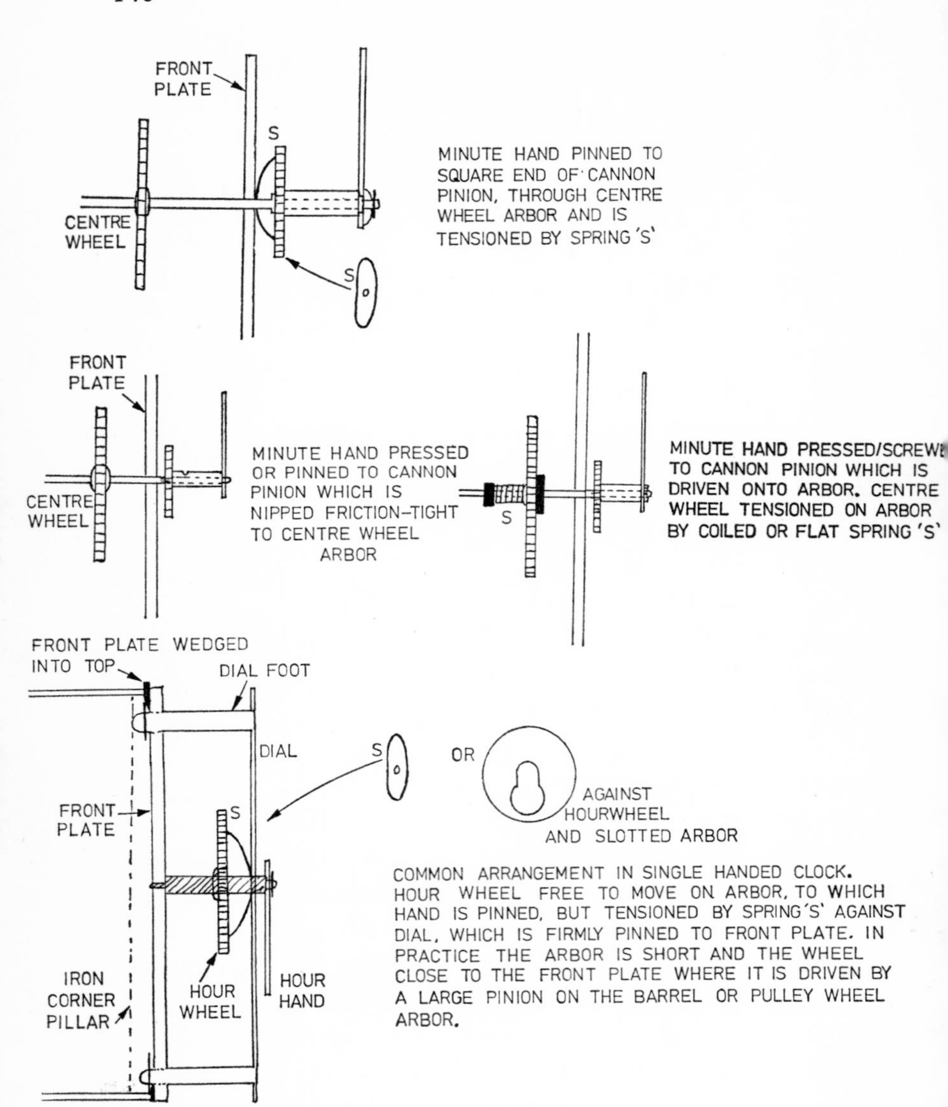

Fig 23 Set hands and cannon pinion arrangements

screwed into the backplate, but in the 'birdcage' movements the bracket is usually mounted on the brass top of the frame. In this latter type of movement the pallet arbor runs from front to back across most of the clock and is pivoted into raised angle-brackets (see Fig 12). In the higher grade clocks of many types, but especially French movements, you will find that certain wheels of the striking train are pivoted into small brass cocks screwed over larger holes in the plates; this arrangement is a great advantage when it comes to setting up the clock to strike correctly, as we shall see later.

Going train

The going train of an eight-day clock has of course more wheels than that of a thirty-hour clock. The number of teeth on the wheels varies in conjunction with the number of leaves on the pinions, but certain conventions exist in the gearing of these trains, especially those with seconds pendulums, and the detail of a missing item can usually be found from a handbook's tables of trains—or it can be calculated starting from the fact that the centre wheel, on which the cannon pinion is fitted, must revolve once in an hour. The basic formula is that the ratio of the wheel teeth to the pinion leaves in the train is the number of revolutions which the fastest wheel (the scapewheel) makes for one revolution of the slowest wheel. The number of teeth on the barrel or fusee, on any intermediate wheel, and the number of leaves on the centre pinion are relevant not to the time-keeping but to how long the clock will run at one winding. The exception is the one-handed clock where the hand's motion wheel is driven directly by the leaves on the pulley-wheel's pinion; here there is no centre wheel as such and all the teeth and leaves have to be included in the calculation.

Click-work

As the hands must be able to rotate free of the relatively

fixed wheels in the gear train, so must the winding barrel and square (or rope pulley), or fusee arbor be able to be turned for winding. This is accomplished by having a ratchet wheel fixed to the winding arbor and engaging, by a sprung pawl, with the first wheel of the train, which will be the base of the fusee or, the wheel of the barrel. In the 'bird cage' and many other early movements there is the same principle, but here the spokes of the pulley-wheel, rather than the teeth of a ratchet, are caught by the pawl (see Fig 24). The pawl and its spring are known as the 'click' and 'click spring', and it goes almost without saying that they take a great deal of strain and must be checked for wear (which can usually be filed out) and firmness of mounting. The circular click spring of the thirty-hour clock frequently needs re-riveting if it is to be secure. In superior long-case clocks you may come across a detent finger resting in a ratchet wheel on the barrel. The pressure of this provides 'maintaining power', keeping the high-grade movement ticking over whilst it is being wound (when all the power of the weight is momentarily removed and indeed, if the click is at all stiff, the wheels may even be trying to run backwards).

These are the main parts of the movement (though others peculiar to spring-driven clocks remain to be considered) and, before you dismantle it, have a good look at them, finding out first where everything is and how it works, and secondly what is clearly going to need repair or replacement once you have it in pieces before you.

Study the operation of the striking system, but do not be too concerned at this stage if the right number of blows is not always struck. This is a matter for adjustment in reassembly. Look out especially for a bent rack tail which does not fall reliably onto the snail; for bent lifting pins which cause the strike to go off early or late; for a missing warning pin (you will see a stub or a hole in the wheel next to the fan), and for sloppy holes for the locking wheel and gathering pallet arbors. All these will stop the strike from functioning or will cause it to work unreliably.

Consider the operation of the escapement of a pendulum clock. If the escapement is reasonably adjusted it should be possible to persuade the crutch to swing to and fro with a little pressure on one of the train wheels. If it swings, but

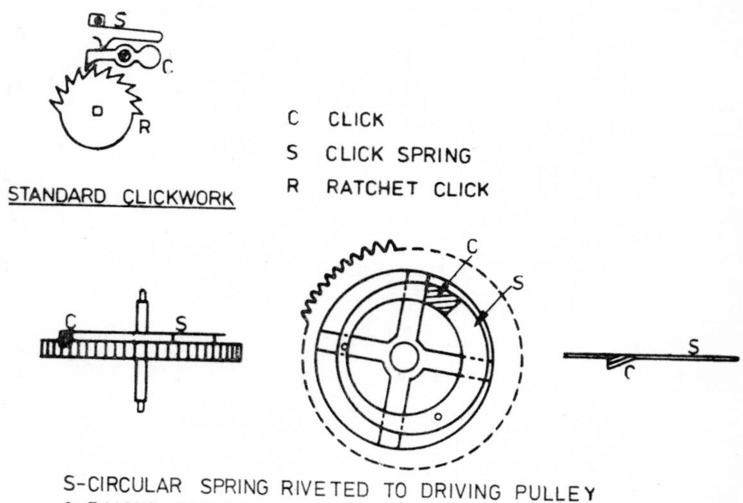

Fig 24 Standard and thirty-hour click-work

very unevenly, try altering the angle of the crutch laterally by bending it slightly until the 'ticks' are evenly spaced. If the whole escapement is jammed up, loosen the suspension cock and see if it can be moved up slightly to release the escapement. If, on the other hand, the scapewheel misses out some teeth or runs round wildly, try lowering the pallets by moving this cock slightly downwards. The correct height of the pallets, the depth of the escapement, is important if the clock is to run to time. Another critical freedom is 'drop'—

the space between the release of one tooth by a pallet and the locking of the next tooth by the other pallet. While you are testing the escapement, you may notice that the pallets are broken or their acting surfaces badly worn. This will have to be remedied. These adjustments and repairs are considered in the next chapter.

Not only in the escapement, but also in all pivot holes, some freedom is essential, but the minimum possible for free action is the ideal condition. Holes which have become widened can best be identified while the movement is still assembled and the pivots are slopping about in them. Lateral freedom is not the only thing to be gauged; arbors also need 'end-shake', that is, freedom from end to end between the plates but, again, not too much. An arbor whose pivot shoulders are hard up against the plates will not hold oil and in due course will bind. An arbor which has too much end-shake, on the other hand, will cause roughness in the action of the gearing, owing to varying surfaces being engaged, and may lead, on the striking side, to imperfect locking or warning of the train.

In brief, you cannot be too thorough in your examination of the clock at this stage. It will help you in assembling it later. Remember that when you have put it together, there is nothing more annoying than a still undetected fault. You will very probably have to strip the whole movement down again, and inevitably there will then be finger-marks to wipe out. Once you have done your inspection, it is not a bad idea to make a sketch of the layout, especially if you are new to the particular type of clock. Finally, if it is a striking clock it is wise to mark the motion wheels at the points where they engage, though sometimes they will already have been marked with punched dots; the correct positioning of these wheels is essential for proper and reliable striking. Then you can start the stripping itself.

In stripping, an orderly procedure is wise. Take off the

crutch and pallets at an early stage—they are easily damaged by laying the movement on its back to work on it. Then unpin the motion and striking work which are on the front plate. It is a good idea to take the parts off, clean them, and lay them out in the order in which they were removed. That way you avoid leaving anything out. Even when you know your way around, it is all too easy to omit a simple part when reassembling; often, for example, it is the spring which acts on the hammer, and this usually has to be screwed on the inside of the plate before the plates are put together. In most striking clocks there is an element of the three-dimensional jig-saw—you may know very well where everything should end up, but if you do not put it back in the right order you may not be able to get it there at all without starting again. When you have removed the external pieces you can unpin the plates and take them apart. Do so gently for two reasons; the first is that it is all too easy to break a fine pivot as you take the top plate away, and the second is that there is always more inside than you think there is and it is as well to see how everything fits into place between the plates.

Many clock wheels look much like each other and, when everything is away from the plates, you may well wonder where on earth you will start when it comes to putting them back. Besides laying the parts out in order, remove the bigger wheels and barrels (as far as possible) next to last and the centre wheel last of all, because these items usually have to go back first. It is also worthwhile using an old box with holes in the top into which the arbors of cleaned wheels are dropped—then you do not have to waste time later wondering if a particular wheel should be 'pinion up' or 'pinion down' in the reassembled movement.

When you have a movement stripped down, try the fit of the plates together. If there has been any distortion one or more pillars may need bending slightly so that the plate drops readily on top. Make any such adjustments now, because they may affect work to be done on the pivot holes, and try the trains in place to see that all is still aligned. Besides being out of true, a movement with ill-fitting plates can be very difficult to assemble and is a likely cause of broken pivots.

STRIPPING SPRING-DRIVEN CLOCKS

Letting down the main spring

For a spring-driven clock the same examination and stripping procedure are generally followed, but there are important differences. The first thing you must do when you have the dial off is to let the mainsprings down. When you bring in a spring-driven clock which has stopped, do not assume that there is no power left in the spring. It may have stopped fully wound and, even if unwound, there will still be some power in the spring unless it is broken. Remember that the fusee system is always set up at a tension even when all the fusee is unwound. If you are worried for the safety of the platform escapement of a carriage or similar clock, wedge the train and then remove it. It is possible to let the train run itself down, but with an eight-day movement it can take a long time and there is the risk of damage, especially to the pivots and striking mechanism, from running at this unnatural speed.

Letting down a powerful mainspring is perhaps the most tense moment in the whole undertaking. You must avoid breaking either your wrist or the movement. If you lose control of the spring, you will almost certainly break one of the larger pinion leaves and strip several of the wheel teeth; repairs to these are frequently beyond your personal scope and the nature of the 'accident' is humiliatingly apparent to the professional who is given the wreckage to repair. It is no use grasping the winding square, or the square of the spring's barrel, in a fusee clock, with big pliers and trusting to luck that they will not slip. You need a large key (not necessarily the one which came with the clock) which is a good close fit to the square, or a stout hand-vice. Turn the key or vice slightly as if to wind up, until the strain of the spring is taken, and then release the click, letting it slide over the ratchet teeth one by one. When your hand can turn no further with proper control, let the click go back into the clickwheel and start again. Slow and steady wins this particular race; the damage of an accident is instantaneous and appalling. Make sure all the springs (including the innocuous-looking alarm spring) are really let down; if you remove the plate of a partly wound clock the pieces fly out in all directions—some may be damaged and you will not be able to inspect in a leisurely manner where they all belong.

Removing the spring

Once you have the top plate off and the wheels out, you will have to dismantle the sub-assemblies. The first job is to remove the spring from its barrel. The top is a brass disc with a slot in it, and it is snapped tightly into the body of the barrel. You can remove the top by inserting and twisting a screwdriver or a piece of wood in the slot, or you can knock the arbor on the bench from the back, which will cause the top to jump out. Neither method is recommended—you may scratch or bend the brass with the screwdriver or distort the barrel and spring by the tap—but you will have to choose between them. The best method of extracting the spring itself is to take hold of the inside coil with pliers, preferably brass-faced or covered with a cloth and then to edge the spring out turn by turn. The alternative is to take hold of the inner end in the pliers at arms' length and pull. In practice one does a bit of both. With a powerful spring it is almost impossible to let the final coils out gradually, although releasing them suddenly is liable to distort the spring. What has to be remembered is that there is a risk, in pulling the spring out, that it will not again lie down flat. If some of the coils rub on the barrel top, the spring's action will be impaired and it will be liable to break. Moreover, in bad cases, the barrel top will be forced off which will prevent the barrel from running true. Therefore do as little pulling and as much easing as possible. Also, if you are faced with a really large spring, say more than half an inch deep, hold the barrel in a stout piece of cloth and if possible do the job inside a bag or a sack; the final flick of a spring coming out can give you a badly cut hand and, if you are obliged to let go, a flying spring barrel is a dangerous object. As a last resort with a large spring remember that if your relations with the material supplier are good he will usually remove and replace the spring for you using his spring winder.

Stop-work

In many foreign clocks of good quality there is 'stop-work' fitted to the going barrel. This is a simple device to prevent both overwinding and the use of the weak end of the spring's run. Like the fusee, it is set up at a tension. You will recognise stop-work by the presence of a small steel wheel, shaped rather like a maltese cross, screwed to the barrel and, on the

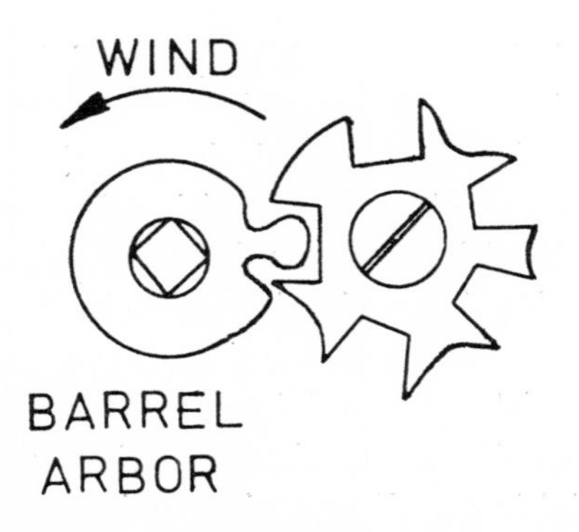

Fig 25 Star-wheel stop-work

barrel arbor, there will be a steel finger with a rounded end (see Fig 25). The finger turns the cross at each rotation until it catches on the one eccentrically shaped tooth of the wheel; usually four or five turns are permitted. Before you take off the barrel top you have to release the tension in the stop-work. Grasp the winding square in a fixed vice and, holding the barrel in a cloth, take up the tension before unscrewing the steel wheel and removing it. Then let the rest of the spring unwind slowly as in the initial unwinding process. You will also find crude stop-work on wooden-framed Dutch clocks and many spring-driven cuckoo clocks. Here a pinion on the barrel arbor engages, outside the plate, with a wheel which has two teeth joined together as a solid block; when the pinion runs into this block, the spring can be wound no further. The system is simple enough but again make sure that the tension is off before you dismantle it.

Fusees

The stop-work of the fusee movement is not a luxury but an essential. It is there to prevent the overwinding of the fusee which would cause the gut or wire to break and let loose the full power of the mainspring with catastrophic results. This stop-work can usually be removed and adjusted after the assembly of the movement and it can be taken off before you separate the plates if you wish. Before you remove it, however, check that the stop-finger, which is spring-mounted behind the plate, really is raised by the gut so that it comes into the path of the index piece (the 'poke') at the top of the fusee to stop the latter from revolving. Take the opportunity also to make an inspection of the condition of the gut or wire. A gut in good condition should be wiped with an oily rag between the fingers to keep it supple. Fusee chains, on the higher-grade pieces, rarely break, but they can get stiff with rust in the joints. They can be cleaned by immersing in petrol, then drying off and wiping with oil. In really bad cases they can first be scraped clean by pulling to and fro across an emery stick.

It is always worth stripping down a fusee, wiping it clean inside, and giving it oil. The base is secured by a spring clip and screw or blind pin. Inside, you will find the click-work proper (for the click and ratchet on the spring barrel have no click spring and are only moved when setting up the spring or letting it down; their function is to keep the arbor still against the power of the unwinding spring). You must make sure here that the click is securely riveted to its base, but free to move, and that the click spring acts freely. In superior movements, this base will itself be a many-toothed ratchet

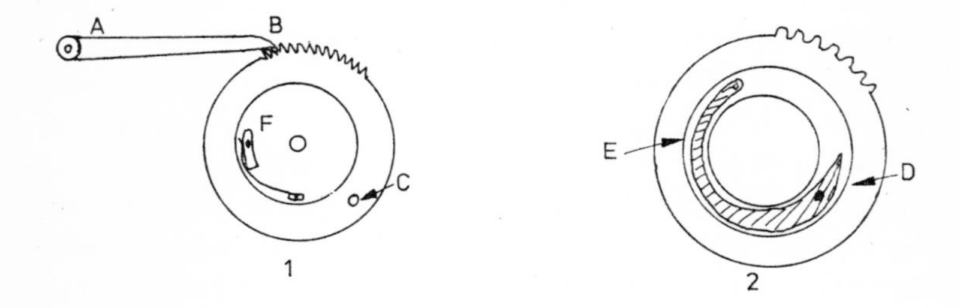

THE MAINTAINING RATCHET (1) WHEEL FITS OVER THE GREATWHEEL (2), THE POST 'D' OF THE MAINTAINING SPRING LOCATING IN HOLE 'C'. IN OPERATION THE RATCHET IS HELD BY THE DETENT 'A' AND THE TRAIN REVOLVES UNDER THE TENSION OF SPRING 'E'. 'F' IS THE MAIN FUSEE CLICK WORKING ON A RATCHET SCREWED TO THE FUSEE.

Fig 26 Fusee maintaining power

wheel engaging with a stout lug on a circular spring which is fixed into the bottom of the fusee (see Fig 26). This is the maintaining power (the purpose being the same as in weight-driven clocks) and the spring is tensioned, when the clock is being wound, by a finger, a long pawl, falling by its own weight and catching in a tooth of this ratchet.

Pendulum suspension

The pendulum suspensions of English 'bracket' and mantel clocks are generally the same as on long-case movements, though you will sometimes find movements in which the pendulum is suspended from a rocking arm, moved up and down by a regulating hand on the front of the dial, the suspension spring moving between close-fitting chops. This is simply a way of regulating the pendulum by altering the effective length of the suspension spring. Other than making sure that the brass chops neither bite on the spring nor permit it to wobble from side to side, no work is usually required here.

The pendulums of French clocks are often regulated by a small square on the dial. This square has a little toothed wheel at the other end of the arbor which engages with a horizontal toothed wheel in the suspension cock assembly which is rather complicated. There are two main types (see Fig 19). In one the turning of the toothed wheel causes small brass chops to rise or fall along the spring, and in the others, the older pattern, the spring itself rises and falls between fixed chops. Other than that these are small parts which tend to require a good deal of cleaning, the servicing is usually confined to the fit of the brass chops, as in the English system. It has to be repeated, however, that French suspension springs are delicate and a bent or broken one, especially in these 'Brocot' regulating devices, should always be replaced.

French clocks

French clocks require special care, but are correspondingly rewarding. They run on very little power, have delicate pivots, and are always highly finished. They are usually held in their cases by two long screws into brass lugs, usually riveted onto the front bezel, which clamp front and back together. In most instances the bell will have to be removed as the first step, but if there is a wire gong the pendulum can be extracted through its centre. When you have the pendulum off and the movement exposed to view, pay particular attention to the striking mechanism. Whichever system it follows, the tolerances here are fine and it is folly to start bending lifting pieces, racks, and the like, until you are absolutely certain that their being out of the true is really the cause of trouble. If the clock is striking incorrectly, but the rack tail falls properly onto the steps of the hour snail, you may well find that the hammer is raised well off the bell or gong when the clock stops striking. If this is the case, the setting-up of the clock has been done incorrectly and you need not fear the need for elaborate repairs or adjustments to the parts. While you are studying the action you will notice that the fan and the pallets are usually pivoted in blocks pushed into the front plate and with either a slot across them or a lug. The pivot holes are eccentrically placed in these blocks and the slot or lug allows the hole to be moved with screwdriver or pliers to adjust the depth of the fly or escapement. Be wary in adjusting these, now or later—they are a quick way to break or bend pivots.

It is particularly important with these clocks that the forked crutch is a close, but yet a friction-free, fit to the pendulum. Make sure as far as possible that the sides of the fork are parallel, vertically and laterally, and polish the inside of the fork with a very fine blunt file, or a burnisher. To close the crutch, insert a rod—an old arbor will do—very slightly thicker than the pendulum rod and lightly tap the fork together. If the gap is very large, solder in a slip of brass. If the fit is tight, file out the slot. The crutch sides must be parallel. The crutch of a French clock is often only friction-tight on the pallet arbor—make sure, however, that it is not sloppy. If it is so the pallets can easily be brought into the right position on the arbor when the clock is set 'in beat'. A loose fit here will put the clock hopelessly out of beat and a very tight fit may lead to damage to the escapement and its pivots. This question of 'beat' is dealt with more fully in the final chapter.

The procedure for stripping is as for other spring-driven clocks, but you may meet complications in dismantling the dial assembly if it is in several parts, especially if there is a visible escapement in front of the dial. Here you will have to remove part of the dial, then the pallets, and then the remainder of the dial. Be careful in removing the gathering pallet. It is a press fit onto a rather delicate arbor. Either prise it off with great care or, preferably, tap the arbor out when the rest of the movement has been stripped. When you come to unpinning, proceed very warily, lifting the plate off slowly in a dead vertical direction, for these pivots are easily bent and snapped. All the surfaces of such movements require to be well-polished, to cut down friction and for appearance's sake.

Carriage clocks

These remarks as to care and finish apply equally to that specialised French clock, the carriage clock, where of course the movement is fully visible. The way into a carriage clock is through the base. First take off the cover, if present. Then remove the four screws, one in each corner, which secure the case pillars to the base. Holding the clock upside down, lift the base and movement out of the rest of the case. In the great majority of these clocks the glass panels, which slide in slots in the pillars, and the door, are not secured in any way, so they must either be held in place by hand or the movement must be taken out upside down in this uncouth manner. In some of the better-quality French clocks, and many of the English ones, however, there is an additional brass base ring screwed into the case pillars and holding the panels and door in place.

The detailed structure of the upper parts of the cases varies (see Fig 27). In the older French clocks the frame is in a single casting and the whole top is glass, secured by a square bezel screwed or pinned to the top of the frame from behind. The handle is sprung between two upright lugs which are part of the casting. In the more standard versions the four

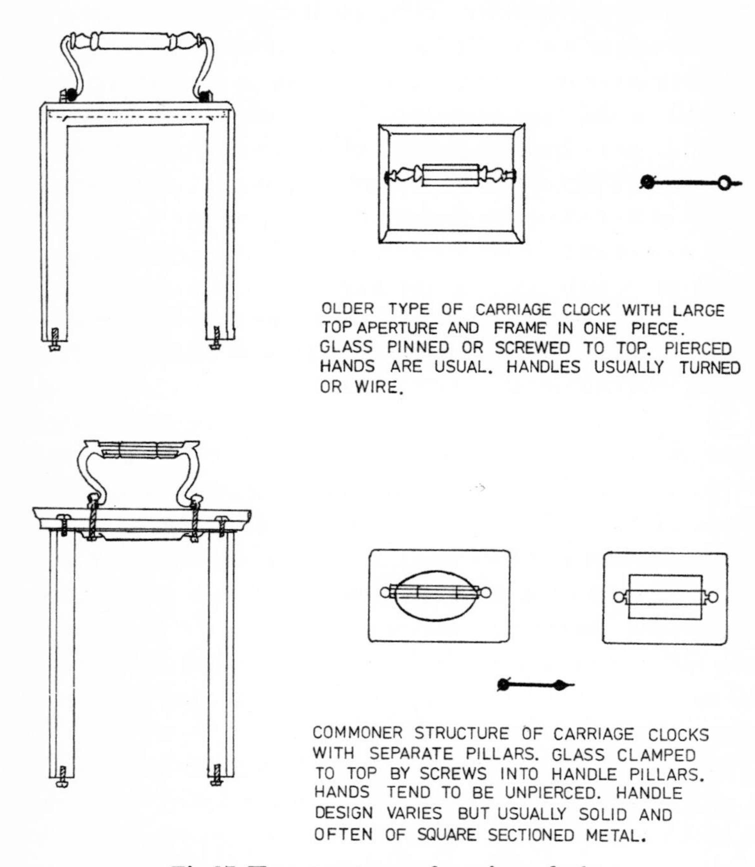

Fig 27 Top structures of carriage clocks

pillars screw into the top separately from above and the fittings are covered by a brass top-plate with a rectangular or oval aperture through which the escapement is seen. The top glass is held in position by clamps whose screws pass upwards through the top and into the small turned pillars between which the carrying handle is sprung. The door is hinged merely on pins turning in holes in the top and base. These cases look best gilt, and you may be able to have the job re-done. A gilt case will often clean up if carefully washed with soap and warm water. Do not use metal polish or you will ruin what remains of the gilt. If the gilt is too patchy to be tolerated, or the case has never been gilded, the alternative treatment is extremely thorough cleaning and polishing and then a thin coat of lacquer brushed on, or, for a pleasant semi-gilt appearance, sprayed from an aerosol.

The movement of a carriage clock is best removed from its base before the dial is unpinned because sometimes the dial is pressed rather tightly against the base and may be damaged if it is removed while the base is still fixed. The movement is usually secured by two large-headed screws passing up through the base and into the two lower plate pillars. In the older movements, there may instead be four small screws through the base and into the plates themselves. The seconds (if fitted) and alarm hand are a press fit—remove them by very gentle leverage from a fine screwdriver placed on a slip of wood or card to protect the dial, or with fine nippers. The minute hand is a press fit on the later clocks but is pinned on the older and better quality movements. The hour hand is invariably pressed on and again care is necessary to avoid damaging the dial.

The older rack-striking carriage clocks tend to have a striking system similar to that of ordinary French striking clocks, but it is more usual in the hey-day of the carriage clock to have a 'flirt' release rather than the conventional lifting piece arrangement, and in this system there is no warning (see Fig 28). If there were a warning the repeating mechanism would be unable to operate for some minutes just before the hour, because the whole striking train would be arrested and awaiting release. The 'flirt' is an angled arm in two parts which are held at the required angle by a delicate spring. The whole arm also moves against a spring mounted on the plate. The tip of the flirt is pushed back by the lifting pins as the cannon pinion rotates. Exactly on the hour the pin passes beyond the tip and the flirt flies across the clock under the power of its spring and knocks the rack-hook away from the rack. The rack then falls and the gathering pallet raises it step by step as the clock strikes in the normal way. A modified form of this system is often also found in modern chiming clocks. It never ceases to amaze me that this Heath Robinson contraption works—but work it does, and reliably, and its proper adjustment is merely a matter of observation and common sense.

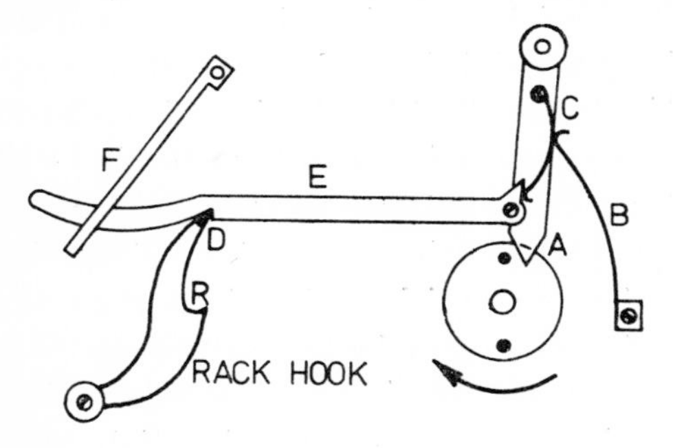

A-LIFTING PIN ABOUT TO DISPLACE FLIRT TO RIGHT
B-SPRING TO ENABLE FLIRT TO RETURN WHEN RELEASED BY LIFTING PIN
C-SPRING HOLDING FLIRT DOWN ONTO PIN D
D-PIN ON RACK HOOK, WHICH WILL BE KNOCKED ASIDE BY FLIRT ONCE RELEASED.
E-FLIRT
F-FORKED ARM WHICH GUIDES FLIRT
R-POINT WHERE RACK IS HELD UNTIL RACKHOOK IS DISPLACED BY FLIRT

Fig 28 Flirt-striking release system

Many carriage clocks, particularly repeaters, are fitted with an alarm. This is, in effect, a very simple and robust anchor escapement with the striking hammer as the pendulum, and it is powered by a separate small mainspring. The hammer usually strikes one of the ordinary striking gongs or bells. The working is simple and reliable, depending on the placing of a small cam on the set-square arbor which, against a spring, allows the arbor and a disc with lowered edge fixed to it to move and release the alarm at the proper time. There is normally no provision for silencing the alarm; if it is wound up and released it will go on sounding until the spring has run down. Have a look at the mechanism, and try it out before you strip the movement, and you will not encounter any problems.

Carriage clocks were so called because they were very much more portable than the pendulum-controlled French clocks of the time and, perhaps because they were, in fact carried on journeys in special cases with rising fronts so that they could be heard while the traveller was in his coach. The appearance of the front plate behind the dial, especially with a full quarter-repeater complete with alarm, can be somewhat daunting. There are, certainly, a great many parts working in the small space and you must take care not to bend any of them until you are positive beyond doubt that they really are distorted. But much of this apparent complication is due to the fact that most of the moving parts which are not actually wheels have springs working on them to keep them in place, and, in addition, the older clocks often have the click-work on the front plate as well. A properly set-up carriage clock will often strike when it is upside down (though there is no requirement for it to take, or to pass, such a test), and certainly changes in position should not interfere with its general running. Time spent on a thorough study of the clock's workings is never wasted and a diagram of the layout will also be useful. Then follow the general rules for stripping —lay out the parts strictly in the order in which they are taken off and keep each part and its associated screws and springs together. Screws and the screwed studs to which flirt, star-wheel, and other parts are secured often appear interchangeable, but in practice they are most often not so. Not only must the screw fit its hole, but the stud must permit free movement of the part which rides on it.

Platform escapements can be tested, if they are not obviously ruined, by turning the scapewheel, rubbing a piece of pegwood steadily against the pinion. This will provide enough power for the escapement to function if all is well. To examine further, and to clean, unscrew the balance cock, invert the escapement and with a gentle wriggle allow the cock and balance to fall together onto your hand immediately below. The cock is fitted on steady pins to keep it in its exact position and may require light prising off with a screwdriver. Do not on any account lift the balance wheel away from the cock or let it hang by its own weight; a distorted hairspring will rub on the cock, stop the clock, and is often extremely difficult to straighten. The whole wheel and spring assembly can be removed by pushing the spring's securing stud out from above, undoing the fixing screw if there is one. Alternatively, the spring may be gingerly unpinned and the wheel turned until it is free. Clean the jewelled pivot holes with a blunt peg—the holes are brittle and a sharp peg may stick in a hole and crack it. The cylinder scapewheel is of delicate bright steel construction and a very soft brushing will usually bring it up bright. The same treatment can be applied to the lever and lever scapewheel, which may be of brass. Clean the pivot holes of these parts similarly and give the stripped platform a good polish, unless it has a frosted gilt finish, in which case it should be washed, preferably in benzine, but a mild soap solution will do no harm, and then be well dried.

When the balancewheel is removed, or before it is replaced, the lever escapement can be further tested by again applying slight power to the scapewheel, and touching the end of the lever, which should jump briskly to the other side, releasing a tooth on the wheel. It is of course this 'jump' which imparts an impulse to the balancewheel. The lever platform of a carriage clock is rarely in a bad way, except that it may be dirty and the hairspring bent. It can be cleaned and time be spent on the hairspring with a needle and tweezers before hope is abandoned.

The balance of a cylinder escapement has no great arc (it is in fact prevented by a pin on the rim from making more than a full revolution) and the action, while it sounds crisp, always looks rather sluggish. If a cylinder escapement is giving no trouble, then clean it and otherwise leave it alone but, if it is unreliable, it is better replaced by a new lever platform; the original escapement may be desirable, but a worn or broken cylinder is not, in most cases, worth the time and trouble of repair or replacement, which are hardly within your scope.

The commonest fault with all these escapements is a broken balance-staff pivot. The ham has not the facilities for turning a new staff, nor can he fit a new pivot. If the pivot is at the top, he will try, no doubt, to improvise—bending the cock down, filing a new point on the broken staff, and so on—but it will be to no avail. The ultimate solution, a replacement platform, is in most cases the only practicable one, and its new vigour will be a marvellous relief after fruitless labours.

Four-hundred-day clocks

There are no special points of note in stripping and cleaning 400-day clocks, other than that the first action must be to unship the heavy balance (which is hooked on) and to remove the fragile suspension spring, and the second to let down, with caution, the extremely long and powerful mainspring. The expert advice must be against leaving the spring in the barrel and dropping a little oil onto the coils, because undoubtedly the spring should be extracted and the old sticky oil be wiped off. But still it may be better to leave well alone in this case and to hope for the best, unless, of course, the spring is broken. These large springs are liable to be distorted in the difficult business of extracting and replacing them, and you cannot afford any loss of power arising from the rubbing of the distorted spring on the barrel. You can, if the job must be done properly, buy from the materials dealer a mainspring-winder (which will allow you to remove and replace the spring ready wound to a size which will slide in and out of the barrel), or you can have the spring dealt with by the dealer himself.

7

Minor surgery

You are now in the position of having the entrails of the clock all cleaned and lying, let us hope, set out before you in the order in which they were removed. When you made your survey of the movement, and again as you cleaned the various parts, you will have noted repairs which needed to be carried out and there are very likely others which may not have struck you. In this chapter are listed, alphabetically, a number of routine repairs which are within most hams' scope. Some apply to almost any movement, and others apply to a particular type which is especially likely to come your way. You can read them straight through, or pick out those which are more obviously necessary.

BLUEING STEEL

Steel is blued for ornament, though it may resist rust longer when blued than when not. It is mainly a question of style. The screws of many stout bracket clock and long-case movements look wrong when blued, whilst those of many French clocks are beautifully done and would look out of place were they not. The best rule is to blue when it has been done before, and this applies to steel hands as well as to screws.

You can paint with blue enamel, but the result will appear heavy and dark beside steel which has received the traditional treatment. This consists of getting the steel absolutely clean and bright and then heating it on a brass plate, when it changes colour quickly through shades of brown to purple and then to blue. When the desired blue is reached the metal is plunged immediately into oil.

The essential thing is to keep the metal free of grease or oil before heating it. A hand with finger-prints on it will never blue properly. For screws the easiest holder is a piece of old clock plate with the screws dropped into the holes. For hands, the process is ideally carried out on a small tray filled with brass filings, for these hold the heat and conduct it evenly.

CHAIN CONVERSION

To the investor and collector converting a rope-driven clock to chain drive seems barbaric, and it may seem so to you as well. The fact is, however, that a new chain system is more efficient and that soft clock rope is both hard to obtain and difficult to splice evenly and without a bump in it. The drawbacks of a new chain drive are the noise of winding up and the fact that it is not original.

Pulleys for rope drive are not suitable directly for chain, for the spikes which stop the rope from slipping are too far apart, too sharp, and too short. New pulleys, complete with a chain of the right size, can be obtained. The spikes which catch the chain are much thicker and are closer together.

This is not a job which the amateur can usually do without a lathe, for the arbors, as a rule, have to be turned down to accommodate the new pulleys, though you may be lucky. The turning is, however, of the simplest and you are unlikely, if you have no lathe, not to be able to find someone who will do it for you for a few pence.

The fixed pulley, which drives the strike, is driven onto its arbor as a tight fit and usually a pin is pushed into the end and right through the arbor. The free pulley, which has the click-work and revolves for winding, is a close, but not a tight, fit and is held back against the wheel, with whose spokes the click engages, by a collet or washer and a pin through the arbor. These pins are tight and often rusty—you will need to knock them out with a punch and hammer. The new pulleys are fitted in exactly the same way, with additional washers to take up any spare longitudinal space.

If you have already a chain clock but the chain is missing, or it has stretched so that it is constantly slipping, it is advisable to send one of the pulleys when ordering new chain. The chains are sized in links to the inch, but it is safer to have the pulley and chain fitted for each other than to rely on measurement since old pulleys often do not fit chain sizes exactly and various sizes will need trying.

COUNTWHEELS

On the older movements—particularly long-case but also bracket clocks—perhaps the commonest part missing is the external countwheel. The clock cannot strike without it and the chances of your finding an odd one of the right size are slim indeed. Fortunately, this is a part which you can make without much difficulty, though it may need more than one attempt to shape it exactly so that the clock strikes correctly at every hour.

The countwheels of these clocks are made of solid brass and ride on a stud (which may also be missing, but a piece of brass rod tapped and screwed into the plate will do), or on the extended arbor of a wheel in the striking train of an eight-day clock. If the wheel was on a stud it will have had a gearwheel fixed to it and be driven by a pinion mounted on the end of the driving pulley. In this type, if the countwheel is missing the gearwheel will also be missing and, as the pinion itself is often only held in place by the overlapping countwheel, it also is quite likely not to be there.

For these external gears you can improvise; the ratio is critical, but the gearing need not be particularly fine in action and there is choice in the number of teeth you have on the wheel and pinion. You can file up a rough pinion without difficulty, or you may be able to find an old one from another purpose which will do. Alternatively, you may be faced with the longer job of making up both pinion and wheel.

The required gear ratio is determined by the number of revolutions which the driving wheel, let us say the pulley-wheel, makes in twelve hours. Remember that the system operates by the coincidence of the counting piece fingers falling into the notch on the locking wheel and the slot on the countwheel. The locking wheel revolves once for each blow struck—that is, 78 times in twelve hours. Therefore, the total reduction ratio of the gearing back from the locking-wheel pinion, through the pulley-wheel and its pinion, to the wheel fixed to the countwheel, must be 78: 1. Count up the number of leaves on the locking-wheel pinion, and the teeth on the pulley-wheel, and see how far you get towards this ratio with the existing wheels—usually it is 13: 1. The locking-wheel therefore revolves 78 times and the pulley-wheel 6 times in the twelve hours. The remaining reduction required is therefore 6: 1, and for each leaf of the pinion which drives the countwheel you will need 6 teeth on the wheel fixed to the countwheel. So, with a four-leaved pinion (such low numbers are not uncommon here) you would need a wheel of 24 teeth, with a six-leaved pinion 36 teeth, and so on.

If you have the pinion, you can calculate the size of the missing wheel quite easily. Its radius will be the distance between the centre of the arbor on which the pinion rides and the mounting stud of the countwheel, less the radius of the pinion. You will have to allow for the interaction of the teeth, of course, so assume the radius is this distance less a little more than half the depth of a pinion-leaf. Draw up a circle of this radius and divide it round by the number of teeth required. Then draw a circle inside it (to the extent of the depth of a pinion-leaf) and across both circles join up your divisions with the centre of the circles. Cut along these lines from the outer to the inner circle, clip off the corners and file the teeth, adjusting the shape gradually until the new wheel will run with the pinion. This wheel is invisible in the clock and may be left solid, but of course you can cut out the sectors ('cross it out') to leave spokes if you wish.

If you must make both wheel and pinion, first experiment with pinions having various numbers of leaves; you will have to settle on a number which will leave room for a wheel of the right size to accommodate the number of teeth required for the ratio (see below on Wheels). You will find it easiest to experiment on card or stiff paper.

The size of the countwheel itself is governed by the distance between its centre and the bottom edge of the counting piece when it is down (ie when the other arm is in the notch of the locking-wheel). This distance will be the inside radius of the countwheel. The outside radius in relation to it is not very critical, save that of course the counting piece when it is 'up' must well clear the raised sections of the countwheel; usually these sections are about a quarter of an inch high. Now draw two circles of the required sizes and proceed to divide them up.

You need 78 divisions, one for each blow of the hammer. To get them you will have to divide the circumference of the outside circle (ie radius × 2 × 3.137) by 78, and then measure this distance as accurately as you can 78 times round the circle. Draw in the divisions to the centre and then cut out the sections round the circle (see Fig 29). Between each hour there is one division cut out. For one o'clock there is no raised section. Two o'clock is one division raised, three o'clock is two divisions raised, and so on round to twelve o'clock (eleven divisions raised) and the double-sized slot at one o'clock.

When you have cut out the wheel, pin or rivet it to the gearwheel and mount it on its stud with a collet or washer to prevent the wheel rubbing on the back plate. The stud is usually slotted on two sides and you will need to make a bent brass spring clip to spring into these slots because the countwheel, though it must be free to revolve, must not wobble (see Fig 30).

Test the striking round all the hours. If there is an error, try moving the countwheel round to a different position—the

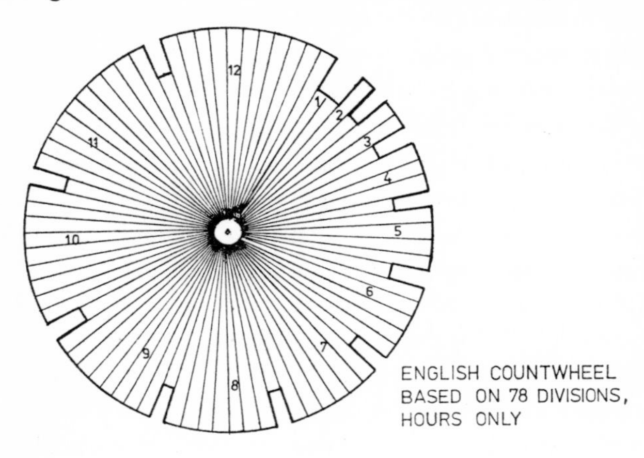

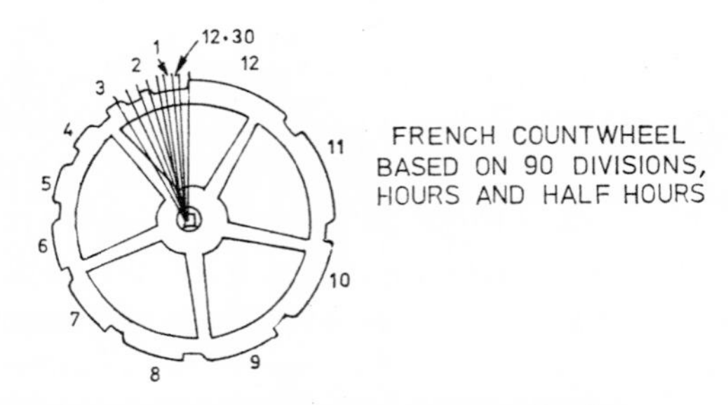

Fig 29 English and French countwheels

exact placing of the wheel and pinion may not be right for this particular countwheel. If, after moving the wheel in relation to the pinion a few times, there are still errors, and they are consistently in the same places, make adjustments to the raised countwheel sections with a file. If, of course, the errors are inconsistent (ie not at the same hours each time round), it is the gearing which is at fault and you will have to check the calculation and remake correctly.

Make a wheel like this out of reasonably heavy sheet metal —about 18 gauge will do. You can do the marking out on a card and stick it to the brass with impact adhesive (which will

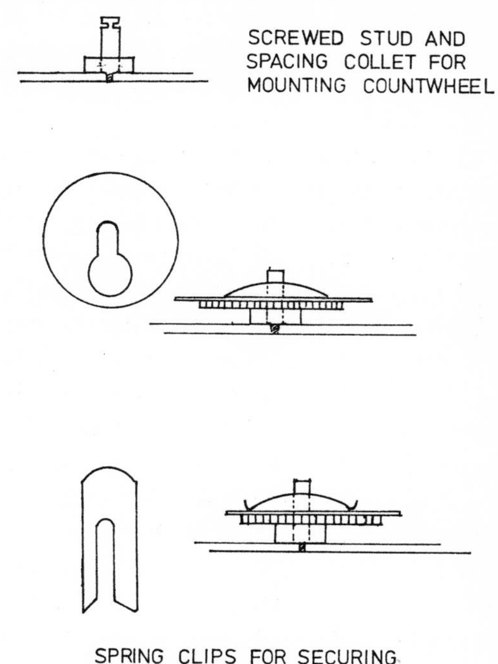

Fig 30 Mounting of English countwheel

peel off when dry). Alternatively, do the job in the traditional manner; rub the brass with a moistened rag charged with copper sulphate, then heat it, when it will turn black and will receive a fine and accurate scratch from a scriber or needle. If the countwheel is none too large and dividing its circumference into this many divisions accurately is difficult, you can draw a very much larger circle round the future wheel, calculate its circumference and divide it similarly, drawing radial lines through to the centre of all the circles; these lines will cross your required circles at the right points.

It should be said that some countwheels on these clocks are of a different type, being a large gearwheel out of which jut segments of various sizes. This type is not easy to make, but can be replaced by the commoner disc type described here. In fact, replacement is less often needed than repair, and the cutting and fitting of a slip of brass to replace a missing segment is not difficult.

Countwheels for French clocks and others striking half hours can be made similarly—though here you will have to take more trouble finishing them and crossing out. To accommodate the half hours you will have to make 90 divisions on the wheel and leave 2 for each gap between the hours (there is no raised section for the single blow at the half hour). In practice the raised sections start at an angle slightly before the beginning of the second divisions between each hour so that the countpiece can ride up easily onto the raised sections (see Fig 29). In such clocks the calculation of the gears and making up of wheel and pinion is not needed, since the countwheel is mounted directly onto the arbor of the second wheel in the striking train.

DIALS

A brass dial with a silvered chapter ring which is in tatty condition can be sent away and made to look like new, if that is what you require. Unless, however, you also want a valuable case made to look like new and propose to have it done by a professional, this may not be the best course. I fancy the general preference is for the figures to be brought out a bit so that the time can be clearly seen, and the silvering to be reasonably restored. Sentiment demands that an ancient clock look cared for but not straight from its maker, and my own feeling is that a fully restored dial is usually an error of taste. On the other hand, a chapter ring which has been sprayed with aluminium paint from an aerosol and the figures filled in with black gloss is an unlovely sight indeed.

The traditional process for silvering a dial is perfectly practicable, although results depend on the quality of the materials available and how good and clean the underlying brass of the chapter ring is. However, a few attempts will usually produce a very reasonable silvered ring.

First remove all traces of the original silvering—you can try without, but the result is liable to be patchy. Do it with emery paper of increasing fineness and try as far as possible to keep the rubbing dead straight in one direction (towards twelve o'clock). Then obtain silver chloride from your chemist and cream of tartar from your general stores (it is used in cooking) and mix the two to a stiff paste. Wear rubber gloves to avoid greasy finger-marks and to protect your hands. If you cannot buy silver chloride (which, incidentally, deteriorates rapidly if exposed to light, so keep it covered when you are not using it), buy silver nitrate instead, either loose or on the ends of sticks sold for removing warts—you will need ten or so such sticks, from which you remove the blobs of nitrate. If you drop the silver nitrate into water (preferably distilled water) and add plenty of salt, silver chloride, which is a whitish purple in colour, will be precipitated and will be usable when you have drained off the water and mixed up a paste with cream of tartar.

Now wet the chapter ring well with strongly salted water and rub it over closely and thoroughly with a moistened rag charged with the silver chloride paste. The brass will turn a disagreeable greyish-brown, but do not worry—this is what is wanted. Mix more cream of tartar into a paste with water and again rub closely over the ring, which will now turn white. Rinse it immediately under very hot water and do not touch the front with the hands. Dry it by waving it in the air or in front of a hair-drier or fan-heater. If, when it is dry, there remain or immediately appear brown streaks on it, you will be able to move them by again rubbing with a stiff cream of tartar paste. The original figures will be hard wax and, if intact, they will come back into condition if you heat the ring slightly. Finally, the ring must be lacquered if it is to stay silver for more than a week or two.

Strictly speaking, the figures are touched up or replaced before silvering. The ring is heated and the deeply engraved figures are filled in with black sealing wax which is chipped to shape when dry and hard. In practice, however, black sealing wax is surprisingly difficult to obtain and if you cannot buy any, oil paint of lamp-black shade can be used after silvering to touch the figures in. Do not use it before silvering; it takes at least a couple of months to dry out thoroughly and it will not withstand your rubbing it with abrasive salt and silver chloride. Apply the paint to the thick figures with a fine paintbrush. Thin it very slightly with linseed oil for the finer lines and use a pen, which is much easier to control.

ESCAPEMENTS

The proper adjustment of an escapement is not a hit-and-miss affair; the exact angles of the locking and impulse faces of the pallets are important and so is their position in relation to the scapewheel. But many of the older anchor escapements were not made with elaborate precision and the finer more modern pendulum escapements do not usually require radical rebuilding to make them work with reasonable efficiency.

The critical factors for our purposes are depth, drop, and beat. 'Depth' and 'drop' concern the relation of the pallets to the wheel. Pallets too close will, extremely, jam against wheel teeth, and, less extremely, produce friction and waste power. From the lateral view, the distance is 'drop', and from the vertical view it is 'depth'.

A pendulum escapement where the surfaces of the pallets have been worn into channels by the dropping of scapewheel teeth onto them will have inexact angles and will also be too shallow—at the worst, teeth will not be locked reliably, and at the best a proper impulse will not be given. There are two solutions—you can move the escapement so that an unworn surface is acted on, or you can repair the pallet surfaces.

You can move onto fresh pallet surfaces either by moving the pallets themselves or by moving the scapewheel. The latter is often easily carried out in the course of routine repairs since the likelihood is that its holes will have to be remade (see below) anyway. If you wish to move the wheel you can fit a bush in such a way that it stands out proud from the inside of the plate and the corresponding holes on the other plate you chamfer out a little to recess it. In the result the whole scapewheel arbor is moved a fraction of an inch between the plates and thus acts on an unworn part of the pallets. Alternatively, if it is a very old clock (say eighteenth century) you will probably find that the scapewheel collet is soft-soldered onto the arbor, the solder will soften when heat is applied, and the wheel itself can be moved. You may be able to do the same with a more recent clock, such as a common French movement, but the wheel will usually have been driven on tightly, the pivots are fine, and it is a risky procedure. Similar methods can be tried on the pallets—but go carefully, for they are very hard and brittle.

To resurface the pallets, heat the tips until soft solder will form a film over them. Heat a piece of spring steel (such as part of an old watch mainspring), but not to red heat, to soften it, brighten it with a file and then tin this with solder and cut off two pieces to cover the pallet faces. Then solder the spring strips to the pallet faces and trim them to shape at the edges with a file. The resulting surfaces will not be as hard as the original pallets, but they will solve the trouble for a while.

I do not myself care for this repair. It does not wear well, it is inclined in due course to mark and score the scapewheel teeth, and, unless it is done very well, it looks untidy. The best course to my mind is to move the escapement, but sometimes you will find that this has already been done, or the pallets are very narrow. Then you must either regrind them (and grinding accurately on a hardened pallet is not easy) or resurface them.

When, by whatever means, the serious wear has been made good or moved out of the way, it may well be found that the depth is too great, teeth are being fouled, or, whilst the clock goes after a fashion, the impulse is too slight to keep the pendulum positively swinging. You are then going to have to alter the depth and, very possibly, close the pallets up. The degree of closure is a matter for experiment—but opening is not so easy, so close up slowly and gently. Place the softened pallets with their back across the jaws of an open vice, which will need protecting with brass chops or hardwood, and then with a stout brass rod (or brass punch if you have one) tap the pallets in the centre.

The depth is increased by drawing the rear pallet hole down or if (as is usual) the pallets pivot into the suspension cock, lowering the cock. You will have to elongate the cock's screwholes upwards slightly. Screw the cock on fairly firmly and then, protecting the brass and with the pallets removed, give it a tap downwards with a light hammer; do not be tempted to remove the steady pins—they will bend sufficiently when you tap on the cock. Making the escapement shallower will hardly be necessary unless you have gone too far, but to do it the procedure is merely reversed.

The commonest of other repairs to pendulum escapements is replacing a worn or missing pallet on the 'visible' escapement at the front of many French clocks. These pallets are exactly semi-circular and they may be of garnet or other semi-precious dark red stone, or of hard steel. Gauge the size carefully from the existing pallet if there is one. The holes, which are filled with shellac, are no guide to size, and too large a pallet will stop the clock whilst one too small will reduce impulse. If you can obtain it, you will not be able to cut the stone, so you will have to replace with steel, which should be driven in but will be satisfactory if pressed into shellac softened under slight heat. Always check that the fitted pallets are upright in their holes.

There are two principles in the positioning of these pallets; the flat sides should be at right-angles to the pallet arbor with the flats facing in the direction in which the wheel rotates, and they should fall on a radial line drawn from the centre of the scapewheel. These escapements are not always perfectly set out, however, and if circumstances force you to a choice, then make sure that the second principle is followed. The scapewheel teeth should drop onto the semi-circle very slightly below its centre-line; as usual, too deep a setting will jam the escapement, and if it is too shallow it will cut down the impulse (see Fig 6).

The wheel teeth of these escapements are very fine and after many years they may become turned over with a hooked appearance. They should be judiciously straightened with the pliers, but do not touch the flat 'front' of the tooth as well for appearances' sake or you will ruin the action.

Finally, of pendulum escapements, it should be noted that the steel pallets of the dead-beat escapement in Vienna regulators are usually clamped into position and reversible, with the unused ends already cut to angle. It is normally only necessary when they are worn to turn them round. These pallets are deceptively simple in appearance. In fact they are cut from a circle of prescribed radius and their correct curve is essential to the action of the escapement. They can of course be adjusted to depth, by moving them in their clamps, and further adjustment is usually possible by turning the turn-table in which the arbor is pivoted, but it is folly to try to bend them. Exactly the same points apply to most older 400-day clock escapements, except that the pallets cannot usually be reversed; they are unlikely, however, to be worn as yet to such an extent that they need to be resurfaced.

I have already suggested that the platform escapement on a carriage clock does not usually arrive in a condition which the ham can do much to improve; either it is satisfactory or you will probably do best to fit a replacement platform (see below). But some adjustments, similar in principle to those for pendulum escapements, can be made to an unhappy lever escapement which is not actually damaged. You can adjust the pallets of a lever for depth and drop if they are jewels pushed into the steel of the lever. They are mounted in shellac and adjustment is carried out with a needle and fine tweezers, resting the lever on a metal block which is gently heated—heat applied direct to the lever itself will be excessive and will not keep the shellac soft long enough for you to work on the pallets.

Remember the tests on platform escapements mentioned in the previous chapter. With the balance removed, cause the scapewheel to revolve and observe the action of the lever. It must be such that the pallets lock each tooth firmly and that neither pallet fouls the tip of a tooth. It must also be such that the lever jumps positively from side to side. If the action is sluggish or faulty, by all means experiment with the pallets, but remember two things—the adjustments required are minute, and what is 'done' in this way to one pallet must be undone to the other, or the last state will be worse than the first. If you move one pallet forward, you must endeavour to push the other back by the same amount. Variation in angle is also possible even with the small slots in which these pallets are mounted. If, when a tooth is just locking, the lever does not swing hard onto the banking pin, the 'draw' is insufficient, then that pallet will need to be turned slightly inwards (it was the absence of an allowance for 'draw' which was possibly the factor, already mentioned, causing the delay in the general adoption of this escapement in the nineteenth century). On the other hand, there must be freedom—the 'run to banking' —after the moment of locking. If the tooth locks and there is no room for the lever to move any more, the banking pin requires adjustment and should ideally be bent in two directions, so that it still presents a vertical surface to the lever after it has been bent.

The final adjustment to any escapement, setting 'in beat', is in some ways the most crucial and certainly the most simple; there are no repair works involved and even an imperfectly repaired escapement may go with increased merriment once it has been set in beat. This will be dealt with, in conjunction with setting up the movement, in the next chapter.

HOLES

The commonest of all repairs to old clocks is attending to the holes. Of course, if when the clock comes to you it goes and keeps good enough time for you, you can skip this job, but you will be laying up trouble for the future, and the improvement which can be made in the running of a train by going over the pivot holes is well worthwhile. The trouble is that over the years the holes become enlarged, friction increases, and the process gradually snowballs. Not only do the holes wear, but under the pressure of spring or weights the wheels move irregularly to and from each other so that the pinions themselves become worn, there is a great waste of power and, in due course, the clock stops. By then it needs to be virtually remade. I remember taking in a distinguished old cuckoo clock, spring-driven and with, when touched up, a bird of some beauty. To my shame, I did not clean the movement or repair it, being busy with other things, but only cleaned up the visible parts and mended the bellows with leather from a pair of kid gloves. There was hardly a hole in the movement which was not oval in shape, and the strike took place, though reliably, with a horrible grinding noise of worn pinions. In due course this clock did not stop or become irregular, it simply exploded as the barrel teeth no longer meshed with the worn leaves of the next pinion. I do not say it was damaged beyond repair, because it is a cardinal principle of the ham that no clock is beyond repair, but certainly repair was not an economic proposition in time or money.

The extent of permissible freedom for the pivots, whether from side to side or from end to end, is a matter of judgment, and you can gauge it only by sight and by feel. A hole in which the oil visibly surges to and from as you wiggle the arbor definitely has too much shake, and any hole which is oval must be dealt with.

The job of straightening holes is known as 'drawing' them and the job of tightening them up is known as 'putting in new holes'. Holes can be literally tightened up—I have done it, I admit, and perhaps no one who spends any time on clocks has not done it sometimes. Undoubtedly you yourself will do it. The procedure is to use a hollow punch whose opening is conical and larger than the existing hole. It has a sprung central pin which fits into the hole as a guide. When this punch is struck with a hammer it will cut a ring round the hole and push the metal inwards. Alternatively, you can go round the edge of the hole closely with an ordinary centre punch and make deep dots in the metal; this will do the same job less efficiently. The result of both methods is unsightly (you will find the ring of dots in many an old movement) and moreover it is ineffective, for the metal which is pushed up into the holes is only part of the thickness of the plate. As a result the pivot has not its proper support and the 'repair' will have to be done again in a relatively short time. If the hole is elongated, it will not be straightened, and if the 'ring of dots' is used it will moreover be a series of bumps rather than a proper circular hole. There is in truth nothing whatever to commend either method but ease and speed and, unfortunately, these are powerful advocates. We hams must confess to this sort of thing from time to time, but we cannot justify it.

Holes are really closed up by clearing away some of the surrounding metal so that a brass bush can be fitted in and the hole in it be adjusted until it is exactly the right size. They are straightened by choosing the right surrounding metal to remove when the bush is fitted. Bushes can be bought in the form of wire, from which the required length can be broken off, or in assortments of ready-cut sizes. These latter can be fitted by patent pressing gadgets, or with a staking tool—both of which ensure that the hole is perfectly vertical at the first go—or by hand, when judgment and experiment with the wheel between the plates are needed to make sure that the new hole is set true. We shall look at this latter method, for the gadgets are expensive to buy at the outset of clock work.

The essential thing to remember is that the pressure on plates and holes in the movement is outwards rather than inwards. We work mainly from the inside of the plate, or else the new hole will be liable to fall out. Start by selecting a bush with a hole just too small for the pivot, and of an outside diameter greater than that of the worn hole. Using broaches, enlarge the existing hole until the bush will just enter, but not go right through the thickness of the plate. To draw the hole back to its original centre, broach more on the unworn side than on the worn—and if the hole is very oval, you will have to remove a considerable amount of metal with a round file—and select a bush of large outside diameter after making the hole round again. At intervals hold the plate at eye-level with the broach resting in the hole, so ensuring that the excavated hole finishes upright. Then tap the bush in, file smooth with emery and polish. If there was previously excessive 'endshake', leave the bush slightly proud to take up this slack. On the outside of the plate, countersink the new hole—this will form an oil sink. Tap in here with a rounded punch to rivet the end of the new hole. Finally, open up the new pivot hole with a fine broach, working alternately from each side, so that the arbor runs smoothly, but not loosely, in it; try the wheel and arbor with its neighbours between the plates and adjust the hole for a good fit.

MAINSPRINGS

It is pointless to try to repair a spring broken midway and repairing the inside end is difficult—the repair may not last long and the spring is likely to be distorted. However, the most usual point of breakage is at the outside end where the spring is hooked to the side of the barrel. In clock work the fastening is almost always by means of a hole or eye placed over a shaped steel hook riveted into the side of the barrel, or held in position by the tension of the spring in a fusee clock. The repair here is perfectly feasible and consists of slightly softening the very end of the spring with heat, drilling it out and filing or broaching the hole to the appropriate shape for the hook (for a round hole on a square hook will certainly break). If the spring is excessively hard for the drill even when softened, it may be punched and then the top of the raised blister is filed off; better this than to heat the end of the spring excessively. While you are on the job, test that the inside end of the spring hooks firmly onto the arbor. If it hooks, it is unlikely to come adrift when the spring is wound up, but there is nothing more annoying than completely assembling a movement and then finding that the spring slips here after only a couple of turns of the key.

The shaping of the hole to the hook, and the finishing of the end, are important (see Fig 31). I have met a spring where the end has been repaired by drilling and then bolted into the barrel. Inevitably, it broke again. The end of the spring changes position when the spring is wound and a rigid fixing will damage it. The hole must be central—equidistant from either edge of the spring—and the corners of the spring should be rounded off. Too long an end before the eye and square corners will strain the hole and lead to fracture.

Attention should be given to the hook itself. If it is worn, it must be touched up with a file or else replaced. If it is loose, the safest procedure is to make a new hook, which is riveted, or tapped and screwed into the barrel and then the

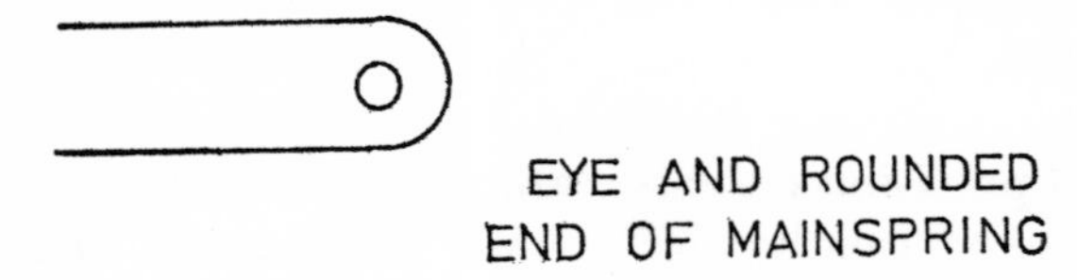

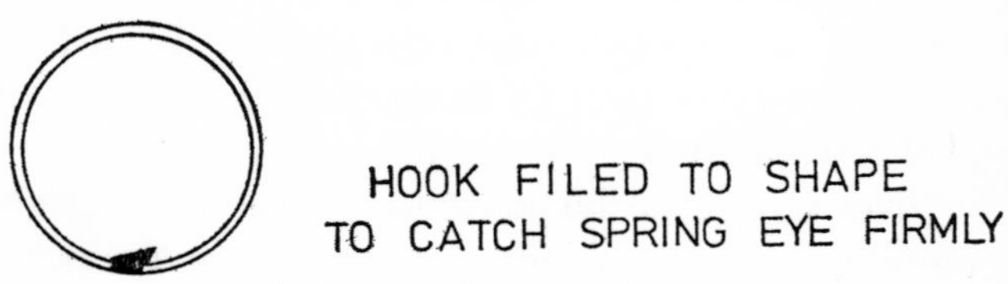

Fig 31 Securing of mainspring

outside end riveted over. Make the hook on the end of strong steel wire, on which then put a good thread. It is possible to make a hook on a long steel screw if you have the right size. Tap the hole in the barrel to the appropriate size, but do not tap it fully. The hook when screwed in from the inside will then cut its own deeper thread and stay firm. Finally, cut off the end and file flush with the outside of the barrel save for the small proudness of the rivet.

Fusee clock springs are often fitted with a square-sectioned hook riveted to the spring and entering at an oblique angle into the thick wall of the barrel—the tension of the spring against this angle holds the hook in place. The replacement work can again be done with a file, but make sure the angled hole is a good fit for the new hook. It is often easier to replace with an eye and a hook fixed to the barrel. Some say that the tip of the spring should be left at a long taper rather than rounded off because this will help to keep it in place. It may; but without a mainspring-winder you will find such a spring very difficult to put into the barrel, and there seems no strong reason why the tip should not be rounded as above.

PIVOTS

Although a precision lathe is really essential for good pivoting, some repairs can be made with reasonable chance of success on the larger arbors without such equipment.

Bent pivots can often be straightened after softening in heat. The safest method is to place a pivot in a brass bush which fits it and then slowly to bend the whole set-up with pliers. 'Slowly' is the operative word; you will need several bends to get the pivot straight from all points of view, and a pivot bent with a good twist to the root will certainly snap off. The pivot can be hardened again by heating to bright red and plunging into water, and then must be tempered, to make it less brittle, by heating much more gently and then plunging it into oil.

On the larger clocks, a rough replacement can be filed up on the arbor, using bushes to protect the arbor and limit the movement of the file. It is best done with a chuck on a motor, but can be worked more slowly in a good quality hand drill mounted in the bench vice. The resulting arbor will of course be short and must be mounted in new holes standing proud from the plates. No craftsman would contemplate such a repair, but the fact remains that it will work if the wheel is not under great strain from a powerful spring.

The more reliable way to replace a pivot (frowned-on one degree less by the professional) is to drill the arbor and insert steel wire of the right size. Again, it is best done with a lathe, but can often be worked without one for a good-sized pivot.

Soften the end of the arbor with heat and then file it flat and mark the centre with a sharp punch. You can buy a jig for holding the arbor and drill in line (and this really is a useful accessory) or you can try without. Drill carefully into the arbor, deeper than the thickness of the plate, with a drill minutely smaller than the pivot-to-be. Cut off the length of steel wire and tap it straight into the hole as far as it will go. Then tidy up with a fine file, flattening the end of the new pivot and slightly rounding off its edges. Make sure that the 'shoulder', where the pivot goes into the arbor, is smooth, and burnish the whole with a fine emery stick and a burnisher if possible. Finally, test the arbor between the plates for endshake, and put in a new hole if needed. This is long, slow work, but it is satisfying if it succeeds, and it is unusual to find more than one broken pivot in a train.

Pivoting in a platform escapement and in the finer wheels and fan of a French or carriage clock is altogether too risky for the inexperienced ham even with a lathe and these jobs should be farmed out to the professional.

PLATFORM ESCAPEMENTS

Replacing a platform escapement is not a difficult job and commonsense is your best guide—with a liberal additive of care, for the new escapement is easily damaged and scratched in the process.

Platforms come in four or five standard sizes. Measure the size of the old one, or space for it, and order the size nearest to it. A perfect fit is not vital and of course the new platform plate can be sawn down and finished with file, emery and burnisher if required.

If the clock is of roughly standard dimensions (say around five inches high in a carriage clock) and has a standard type of lever or cylinder escapement, nine times out of ten the scapewheel pinion will be of eight leaves. If it has, or had, an unusual escapement, with a contrate wheel only about half an inch in diameter (particularly the old lever types where the platform is set very low down in a deep recess in the plates and has or had a helical hairspring), the likelihood is that the pinion should have twelve leaves. The replacement escapement can be supplied with the required pinion, but it *is* essential for the pinion to be of the right size. If, after investigation, you remain in doubt, you must either take the clock to a jeweller or materials dealer and have the point ascertained, or calculate theoretically the size of pinion needed. If the wrong pinion is fitted, it is most unlikely that the escapement will run at all, and the difference in timekeeping produced by using the wrong pinion is too great to be corrected by adding or removing weights on the balance-wheel or by changing the hairspring.

To calculate the number of scapewheel pinion leaves you must work from two known factors—that the centre wheel rotates once in an hour and that the balance-wheel makes a certain number of vibrations in an hour. You can swing the balance, if you have it, and see how many vibrations it makes in a given time and multiply up to give you the vibrations in an hour. It does not follow with an old escapement, however, that the replacement balance will be similar, and in any case you may not have the old balance; then you will have to get the information in advance from the supplier of the platform. Once you know the timing of the balance, you can calculate how many revolutions the scapewheel will make in an hour; modern scapewheels all have fifteen teeth, and you have to allow two vibrations of the balance-wheel for each tooth of the scapewheel. The number of revolutions of the scapewheel to one revolution of the centre wheel (ie in an hour) is the total ratio of gearing needed between those wheels.

Now you have to calculate the ratios of the existing gears, and from this you will be able to discover the number of leaves on the missing pinion which will make up the total required ratio. Count the total number of teeth on the wheels involved. These are the centre wheel, fourth wheel, contrate wheel, and 'x' (the unknown scapewheel pinion). The centre wheel pinion and great wheel pinion are not involved—they are concerned with how long the clock will run for on one wind, not with its time-keeping.

Set the calculation out as a formula, in which the oblique lines represent the gear ratios of the pinions to the wheels. For example, if the balance-wheel vibrated 21,600 times in an hour with a 15-tooth scapewheel, the scapewheel would revolve 720 times an hour (21,600 ÷ (15 × 2)). A suitable train might be as follows;

	Centre Wheel	4th Wheel	Contrate	Scapewheel
Wheel teeth	80	72	64	(15)
Pinion leaves		8	8	x

Here the fourth wheel revolves 10 times an hour (80 ÷ 8) and the contrate wheel revolves 9 times (72 ÷ 8) for each revolution of the fourth wheel, ie 90 times an hour in all. What you have to do is to make this existing ratio of 90: 1 up to the required 720: 1. Divide the required ratio by the existing ratio (720 ÷ 90) and you will have the ratio of the scapewheel pinion (x) to the teeth on the contrate wheel (ie 8: 1). The contrate wheel in the example has 64 teeth and therefore the missing pinion will need 8 leaves to obtain the ratio of 8: 1.

The escapement will be mounted by four screws, usually with raised heads and handsome domed washers. The exact position, which controls the depth of engagement with the contrate gearwheel, is very critical, but must be found by experiment. If you have the original platform, your problems are small. Remove all the parts from both platforms and mark, through the centre of the mounting holes, their position onto the new platform. Punch the marks carefully and then drill them to holes slightly large for the fitting screws.

If you have not the original platform, drop the new one into position, keep hold of it and apply slight power to the train, adjusting the position of the platform until the escapement comes to life. Then scratch the underside of the new platform where the inner edge of the front plate falls. Remove the platform and insert into each hole on the plates a pin of close, but not tight, fit, to stand slightly proud of the plate and dead upright. Onto each pin put a minute drop of blue enamel or similar marker, and drop the platform back, in line with your scratch, on top of the pins, which will then mark where the holes must be made. Dismantle the platform and drill the holes slightly large, as above.

The purpose of drilling the mounting holes slightly large is to allow room for adjustment to that critical contrate-wheel depth. When you have put the escapement in, you will have to see how it runs and adjust this depth for best running; too deep or too shallow will stop the clock. You have also a further adjustment to make. The back pivot of the contrate wheel is mounted in a screw (see Fig 32), or else there is a screw in a separate little cock arranged to press onto the end of this pivot. Screwing in will tighten the wheel until it binds, but you will not be able to adjust this bearing for best running unless the platform is basically in the right position.

If the clock is an old one, it does not follow that the new platform will fit immediately into place. The older escapements, even of conventional design, were often mounted low (conversely, some were mounted flat on top of the plates with no recess at all) and, especially if the contrate wheel is a small one, the bottom cock of the scapewheel in the new platform is liable to foul the contrate-wheel arbor. Strictly, the scapewheel arbor should be shortened and a smaller cock fitted, but this can hardly be done by the ham. Therefore the new platform will have to be mounted higher than the old. The raising can be accomplished crudely by using extra-long mounting screws with bushes between the platform and the edges of the plates or, better, by brass strips screwed into the original fixing holes and with the heads countersunk, or with pins tapped tightly home, the slips being finished well flush with the plates where they meet.

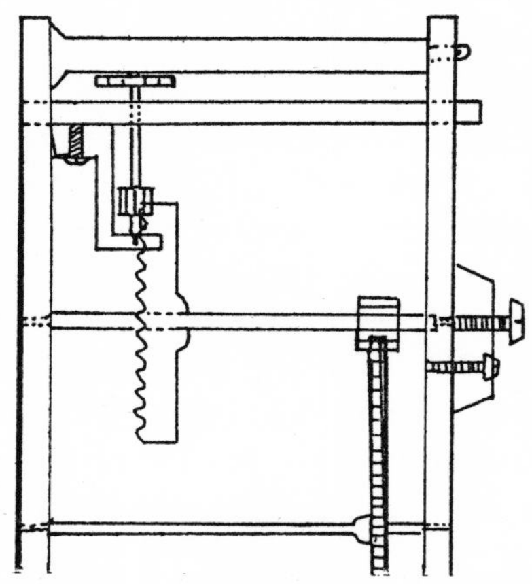

Fig 32 Carriage clock platform, scapewheel and contrate wheel

Again, it may be that with the original-sized pinion but a modern escapement, the clock will gain enormously. Small brass weights can be hooked or screwed to the balance (so long as they clear its cock), or some of the regulating screws round the rim can be replaced, at equal intervals round the wheel, with heavier screws. The hairspring can be changed. If the new platform is reasonably fitted the ham will be able to improvise without great difficulty to get it to keep very fair time and have a good appearance although it may be very different from the original type of escapement.

All new platforms come with a long index jutting out the back. Hold this close to the balance cock with round pliers, taking all strain off the cock, and bend the index at right-angles towards the base of the clock (not too sharp a bend, or the index will snap). It will then go into the case and project downwards to be moved for regulating through the back door.

RIVETING

In the course of your experiments, and particularly when fitting a wheel or hand from its box to an existing arbor or collet, you will frequently require to make a firm riveted joint. This is much tidier and, if done well, stronger than resorting to soft solder.

The technique is to reduce the collet to a size where the hand or wheel is a tight fit over its top edge, and to do so with a very slight bevel so that, when the hand or wheel is right on, it will be a free—but still a good central—fit. Once the rebate has been made in the collet and the remaining shoulder is true and level, the part is pressed or tapped on with a hollow punch against the shoulder. Then the bevelled edge is hammered down over the part with a punch and the rivet filed smooth and flat (but not, of course, filed away). The secret with these round parts is, when hammering the rivet, to keep the collet always rotating; that way, the force of the riveting will be equal all round and, provided the shoulder was carefully made, the part will run true.

WEIGHT AND BOB MAKING

Weights and bobs can be bought, but the finish is not always what is wanted. They can also be made. The weight is as little as you can get away with. For a thirty-hour chain clock this is normally about 8–10lb, depending on how badly the train is worn. With the eight-day clock the going weight will be this or a little more and the striking weight will be slightly heavier, though the lighter it is, the slower and more dignified will be the strike. Pendulum bobs for long-case clocks weigh in the region of 2lb. Bobs in spring-driven clocks are of course much lighter. It has been noted that the bobs, and indeed the whole pendulums, of French clocks are often interchangeable and with these it will rarely be worth making a new bob. The heavier the bob, and the lighter the driving weight which will keep it going, the better.

Driving weights of good appearance can be made of scrap lead from a builder's yard and brass tubing, which in due course can be polished and lacquered. The tube will need to be about two inches in diameter and, filled with lead, will weigh about 8lb if nine inches long. You will need to block the bottom end, and possibly the top as well, with a soldered disc of brass or with a wooden disc which can be painted black for appearance (a two-inch wooden wheel for toys, trolleys etc will do quite well with the edge squared off). Melt the lead, adding bit by bit, in an old saucepan, skim off the dross, and pour the molten metal into the tube. It is easily drilled for a hook or ring to be fitted at the top. A tubular pendulum bob can be made in the same way, with a shorter tube. Pour

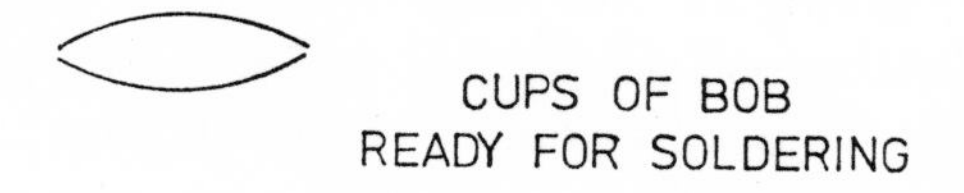

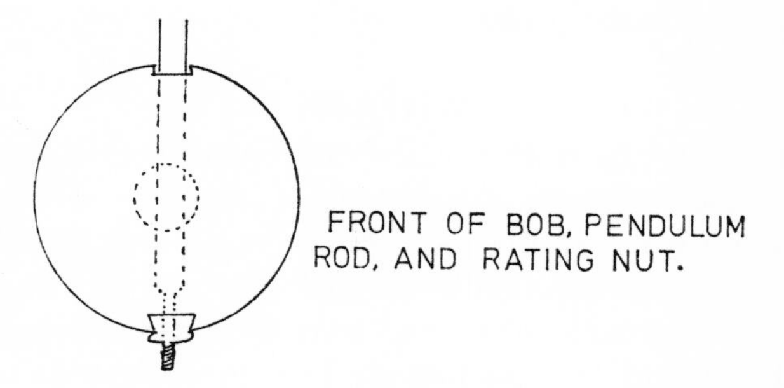

Fig 33 Make-up of pendulum bob

in the lead with the pendulum rod, usually of square-sectioned brass at this end, in position. The rod must, of course, be free to move in the final casting and this is arranged by coating it with oil and graphite; heavy and thorough rubbing with a soft pencil will do.

The 'disc' pendulum bob is made with two discs of brass sheet (see Fig 33). Hollow out a cupped mould in a hardwood block and beat the brass discs with a soft-headed hammer until they are shaped into shallow cups. Tin the edges of the brass cups well with solder, after making sure that they fit closely together all round, and then 'sweat' solder them together by applying heat. (Needless to say, the edges must be perfectly clean if the solder is to 'take'). Now saw slots diametrically opposite each other in the brass case; these are for the pendulum rod, which is again lubricated with graphite to prevent its binding in the lead. In one side (which will be the back) drill a good-sized hole and insert into it a cardboard funnel. Through this pour in the molten lead. Finally, extract the funnel, file off and polish the brass case and the bulb of lead at the back.

WHEELS

Apart from the coarser wheels of long-case clocks and the larger 'bracket' clocks, the making of new gearwheels is not within the scope of the ham without elaborate equipment, and it has been repeatedly stressed that you take a risk in buying a clock with missing or severely damaged wheels and most likely will have to include in your budget the cost of having new wheels cut.

Obviously, this is always the best way from the point of view of lasting efficiency and appearance. The fact remains, however, that many wheels with one or two teeth missing can be reasonably repaired and large brass wheels without too many teeth can sometimes be made. Steel pinions with leaves missing can only be replaced, but 'lantern' pinions, which consist of two brass collets separated by the required number of steel wires as leaves, can be repaired. They were not fitted to the older clocks and not to clocks of high quality, but they work satisfactorily and this facility for repair is a considerable merit.

A single tooth is made from brass wire flattened to the thickness of the wheel. Do the flattening with a hammer, rather than removing metal with a file, because the hammering will harden the brass. Remove the old tooth's remains and file a deep slot into the wheel rim to fit the brass wire closely. Heat the slot slightly and tin it with solder. Tin the pin also and insert it into the slot, heating to solder the joint. Make sure that the wire pin and slot are upright (ie on a radial line from the wheel centre). Finally, file the pin to tooth shape. In this, as in all soldering jobs, file away any surplus solder and be very careful to wash off all visible flux, which is corrosive, when the joint has been made.

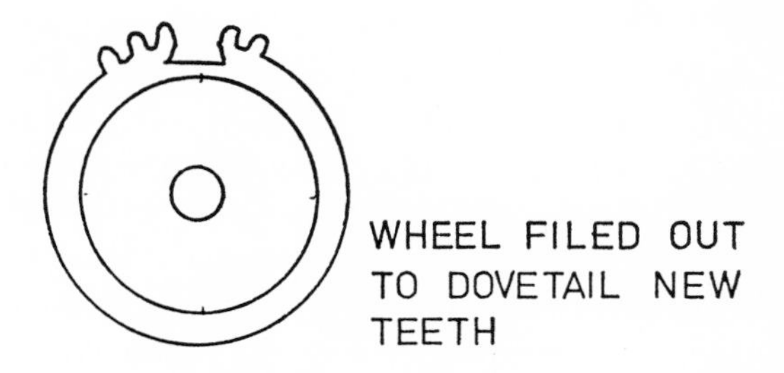

Fig 34 Repairing a wheel

According to the strain on the wheel, two and even three consecutive teeth can be replaced in a similar way with a strip of brass filed to shape and soldered in. When replacing a section of wheel in this way, dovetail the fitting slot; the result will be firmer because there will be a larger area of solder to hold the piece in place (see Fig 34).

Broken teeth on going barrels occur quite often, since a spring rarely breaks when fully wound without leaving such evidence behind it. Moreover, the whole barrel cannot always be seen when the clock is inspected for purchase. Repairs can be made with a good chance of success for one or two teeth, but large broken sections mean that a new barrel will have to be cut. The procedure to repair is to file the remains of the teeth flat and then to drill well down into the barrel base and to drive in brass wire, hammered flat, and as stout as can be managed according to the size of the tooth. If the thickness of the metal permits, the tooth can be lightly threaded and then screwed in tightly. Steel meshing with steel, or brass with brass, is not good practice and will eventually lead to wear, but of course steel can be driven into the barrel more tightly and is stronger, which is important here, and you may well settle for steel teeth regardless of theory and appearance. Finally, the brass or steel is finished off as when repairing a wheel tooth.

If you are supplying a new tooth, let alone a whole new wheel, it is only sensible to try the repaired part together with its neighbour between the plates before you reassemble the movement. It is not easy to judge the exact profile of a tooth visually, and you do not want trouble in the gearing once the movement is back together again.

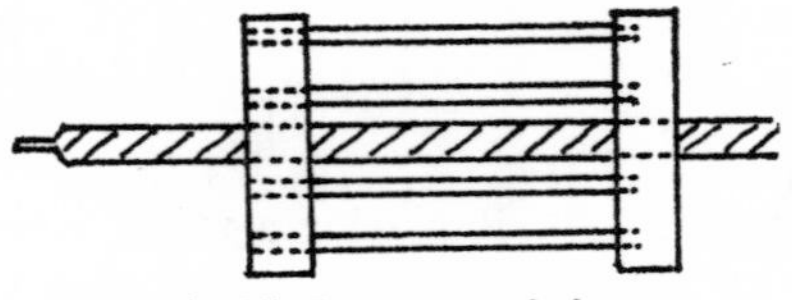

Fig 35 Lantern pinion

The steel pins which form the leaves (trundles) of a lantern pinion are pushed through holes in one brass collet and bedded into the other (see Fig 35). Occasionally the holes in both collets are open, but usually those in one collet are blind. To replace a broken leaf, push out the old leaf with pliers through the open hole. This will be roughly riveted over and you may need to scrape away a little metal to open the hole. Replace with steel wire of exactly the same thickness (this is important) and with a punch tap some of the metal over the hole to secure the leaf (or solder over the top). Try not to move either of the collets, which may only be driven on or may be soldered on the arbor. It is essential that the collets be firm and aligned and you do not want all the leaves falling out during the job or when the clock is wound.

Two points in connection with lantern pinions are worth noting. The first is that a repair is entirely within the ham's scope (whereas the replacement, let alone repair, of a 'solid' pinion is not). The second point is that lantern pinions not perfectly set up do tend to go after a fashion whereas a solid pinion would not. When a pinion leaf, or trundle of a lantern pinion, gives way, the preceding wheel, no longer driving anything, will whirr round at a great pace and in the process will tend to lose the tops of most of its teeth; all the teeth usually remain, but they are stripped to half size. Such a wheel will probably not even meet the repaired pinion, let alone engage with it. Therefore, when you have repaired the pinion, assemble the affected wheel and pinion between the plates. Most probably you will have to attend to the damaged wheel teeth with a file. You want to restore them to something near their original shape, so you gingerly file away the corners, but no more than the corners because the teeth are doubtless already seriously shortened. It is when the short wheel teeth are too far from the pinion to engage securely that one sees advantages in the tolerant nature of lantern pinion gearing. Draw the wheel hole which is nearest the actual gearing, moving it up and then bushing it to the right size. Very likely you will find that with this adjustment you have brought pinion and wheel teeth together again once more. You may have to draw the other hole of the wheel arbor also, but the tolerance of the gearing is such that it will work even though pinion and wheel are not perfectly aligned. Make, by small degrees, what adjustments you can always bearing in mind that you want the depth of the other gearing lessened as little as possible; this is why you make the adjustments to one side at a time, for you want to move the wheel just enough for it to engage with the lantern pinion, but not so far that its engagement with the preceding wheel is affected.

This is not a repair which the professional will regard with pleasure. The fact remains, however, that it works surprisingly well, and most clocks with lantern pinions are not of a value on which many professionals are prepared to spend much time. It is just the sort of job where the ham can operate for his own and others' benefit.

If you are making up a wheel or hoping for one which will do, you need to know its size. The outside radii of the wheel and next pinion added together less the depth of their engagement is, naturally, the distance between their holes. You will know how much to deduct for the existing wheel or pinion. Twice the remainder is the diameter of the missing item at the point of engagement; to it you will need to add a little (for the part of the tooth or leaf which engages) to give you the full outside diameter of the missing wheel or pinion. You can run a check on the basis of the fundamental principle that the size of a wheel is in the same proportion to the pinion as is the number of teeth related to pinion leaves, ie a wheel of 60 teeth driving a pinion of ten leaves must be six times the diameter of the pinion measured at the point where they engage.

The tooth divisions, if not a multiple of four in number, will have to be found by dividing up the circumference by the number of teeth there are to be. The circumference is the diameter multiplied by 3.147. If the number of teeth is devisible by four you can of course successively bisect the circle and resulting sectors until the required number of divisions has been made.

This is not a scientific way of replacing a wheel, but it will, after a little adjustment with the file, often produce adequate results on a relatively large and crude piece. Other wheels will have to be machine-cut.

8

Putting together and setting up

ASSEMBLY

Everything is polished and repairs have been effected. New and repaired parts have been tested and run satisfactorily in the plates. Now it all has to be reassembled, preferably without those infuriating errors which mean another dissection to make good and the spoiling of some of the polished work. What can be done to assist? We shall deal first with points of assembly common to most types of clock, and then with the setting-up procedure for the most frequently met types which differ or need special attention.

First assemble as far as possible the various sub-units. Put together the French suspension, get the spring into its barrel, and fasten the gut or wire to the inside of the barrel and fusee if there is one.

Mainspring and fusee

Unless you have gone to the expense of buying a mainspring-winder you will have to replace the spring by hand. Make sure it is the right way round. In most cases you cannot go wrong because it will only catch on the barrel hook in one direction. Then, with the arbor out, insert the spring's outer end into the barrel by the hook. Let it be a little past the hook because the spring will turn slightly as you fit it and often it will not hook up properly until almost fully in. This is especially true if the end after the eye has been left long. Then, rotating the barrel, push the end of the spring down and gradually wind it in. It is a slow job and very hard on the hands. If you let go before the spring is fully in, it will simply jump out and you will have to start all over again. When all the spring is inside the barrel, make sure the hook has engaged with the eye and insert the arbor, checking again that the inside end engages firmly with the arbor hook. Put a few drops of oil on the edges of the spring and on the two fixing hooks and then put the top on. The barrel cover snaps into the barrel's bevelled edge, but in most cases it will only go in properly as it was originally fitted. Usually there is a punched dot in the barrel wall (sometimes two dots are used to indicate the striking barrel) with which the slot in the barrel cover should be aligned. Snap the top in with the hammer, while protecting the barrel with a brush handle or piece of wood. Now grip the winding square in the vice and wind up the barrel a full turn against the spring to make quite sure that the arbor and spring are really hooked together; if there is any sign of slip, you must take the cover off and correct the curve of the spring's end. It is better to take time now than to have to pull the movement to pieces again later.

The gut or wire of a weight-driven clock is secured by a knot or bow after passing through a single hole in the barrel. Make what sort of knot you like, but cut the end off short, singe the end of gut to stop it fraying and ensure that it is turned in such a way that the gut or wire falls readily into the first groove on the barrel. The gut or wire goes into the fusee of a spring-driven clock in the same way, but in these clocks, with their powerful springs, there are normally three holes in the barrel. There is only one reliable way of securing gut or wire in these holes (see Fig 36). After fastening, touch the end of the gut on red-hot metal; this searing will spread the tip into a mushroom which cannot slip from the knot. If there is a chain rather than gut or wire on the fusee clock, you will have noticed that its end hooks are of a distinctive shape. The simple round hook goes into the fusee, where it curls round a little post in a slot at the bottom. The other hook, shaped roughly like half an anchor, is for the barrel. The chain, however, is not put into place until the movement is assembled.

FUSEE HOOK BARREL HOOK

FUSEE CHAIN HOOKS

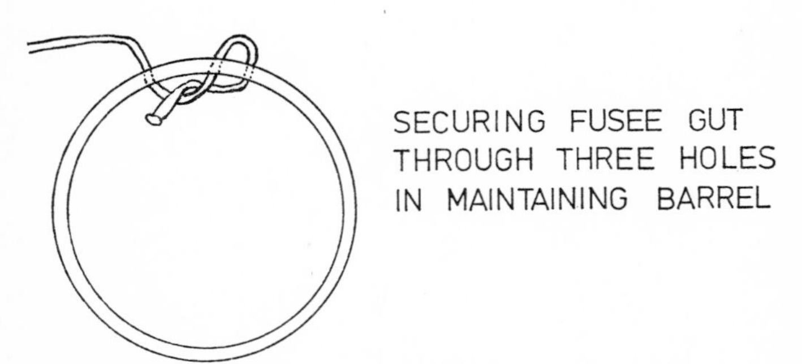

Fig 36 Fusee chain hooks and securing of gut to barrel

If you are dealing with a going barrel which has star-wheel stop-work, you may set it up at this stage (although in many of these movements the barrel is mounted on a detachable bracket and the stop-work can be set up later). The purpose, as explained, is to prevent over-winding and to ensure that the last turn or so of the spring, where the power is much reduced, is not employed. Holding the arbor in the vice and the barrel in your hand, wind the spring up a turn and then fit the star-wheel in such a position that the raised tooth butts against the stop-work finger on the barrel arbor so that the spring cannot unwind, and screw it home securely.

Gear trains

Now you are ready to put the movement together. First screw on all attachments which go onto the inside of the plates (not omitting the vital one already mentioned, the hammer spring for long-case and bracket clocks which often cannot be fitted afterwards). Then place the plates in position, find suitable pins for the pillars, cut them off and file them smooth and flat, level with the edge of the plate when they are pushed home. Assembly is so much easier if you are not fishing around for a pin at the crucial moment.

You put the wheels back in the reverse order to that in which you took them out—this is where an orderly layout during cleaning pays. The best plan is to put the plate with the pillars on top of an open box a little smaller than the plate and to arrange this near to eye-level. This way the projecting arbors will not trouble you as you drop the wheels into their holes and you will not tie yourself in knots as you later concentrate on the difficult job of getting the pivots into the holes of the top plate. Replacing the top plate is never easy and with some movements, especially French ones, it can take a long time and would try the patience of a saint. Start at one corner—that of the going train is best. Get the barrel arbor and the pivots of the first two wheels in, as well as the projections, if any, for the hammer and countwheel, and then you will be able to insert one of the securing pins halfway into the pillar at this corner as a holding device. The other pivots must all go into their holes. It is no use employing force, although light pressure is necessary or the plate will just float around on the ends of half the pivots. Excessive pressure will scratch the inside of the plate and break pivots. In good movements you will often find that the pivots are of roughly graded lengths, the longest ones towards the barrels and the shortest ones at the front on the fan and scapewheel arbors; this makes life comparatively simple. When you have all the pivots in their holes, the top plate will drop by its own weight into position on the pillar shoulders—it is a satisfying sound. Pin the plate loosely and try each of the trains with finger pressure to make sure it is running freely but, if this is a striking clock, do not push the pins right home yet, for there are important adjustments to be made.

Setting up the striking

There are two golden rules in setting up a striking or chiming train. If you forget them, only with the greatest luck will the strike operate satisfactorily and most probably it will be completely unreliable. The two rules are that, when the clock has finished striking, the hammer must be down. In other words, the tail of the hammer or hammers must be completely free of, and preferably half-way between, the pins on the pin-wheel which operate it. Another way of saying the same thing is that, at the moment when the hammer is released and strikes the gong, the locking pin on the locking-wheel must be approaching, but not on, the locking piece which stops the train. An eighth of a turn away is a safe distance. The same applies to the 'hoop' and arm which lock the train in English countwheel striking, and to the tail of the gathering pallet and rack pin in old English rack striking. The second rule is that, where the clock has a warning (and this, as we have seen, is usual except in repeaters), there must be room for the warning to run. Therefore when the clock has just struck, and when the striking train is locked, the warning pin must be at least half a turn of its wheel away from the lifting piece which it engages to hold up the train a few minutes before striking (see Fig 37).

If you are working on a French clock with separate cocks

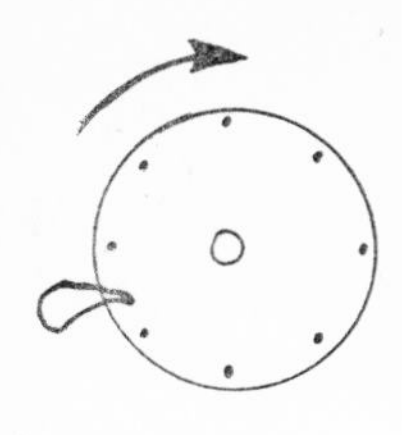

POSITION OF HAMMER TAIL BETWEEN PINS OF HAMMER WHEEL WHEN TRAIN IS LOCKED (all systems).

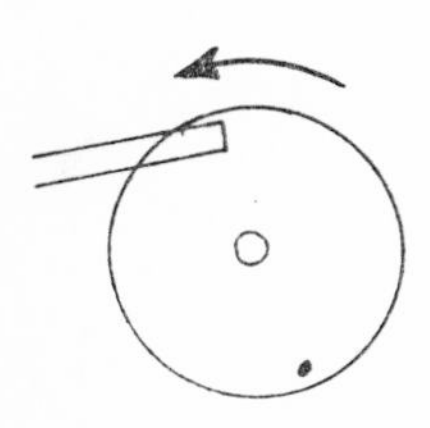

POSITION OF WARNING PIN TO LIFTING PIECE WHEN TRAIN IS LOCKED (all systems which have a warning)

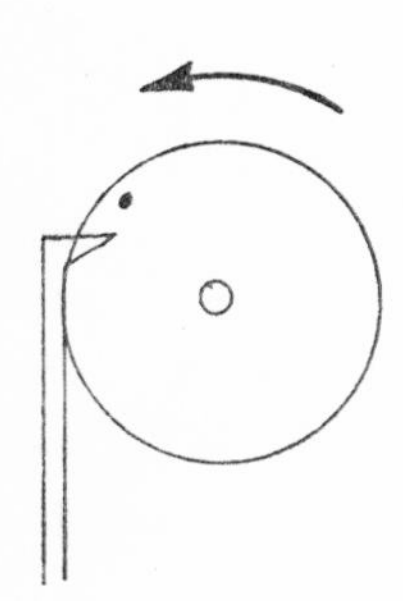

POSITION OF LOCKING PIN TO LOCKING PIECE AS HAMMER STRIKES GONG. (French rack and countwheel systems)

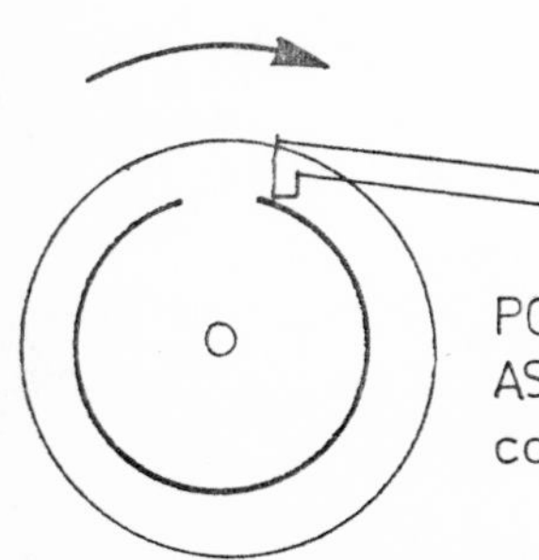

POSITION OF HOOP TO LOCKING PIECE AS HAMMER STRIKES BELL (English countwheel system).

Fig 37 Setting up the striking

for the wheels concerned, you can, of course, at this stage lift the wheels and turn them to the right positions. Otherwise, you will have to raise the top plate on the striking side (loosening the other side), hold the pinwheel still with the hammer between the pins, and juggle the locking and warning wheels until they are correct. It is a fiddly job, but the smooth and positive action of the striking depends on its being properly done.

Now push the pins home in their pillars and proceed to fit the external parts to the back plate. Go on to replace any screwed studs and detent springs onto the front plate before putting on the lifting piece, rack, flirt, and motion wheels. As a rule these parts have to be replaced in a certain order. For example, if you put the hour wheel on too early you will almost certainly have to take it off again to fit the lifting piece. When you push on the gathering pallet of a French clock, make sure it is the right way round to catch up the teeth on the rack; usually the hole through it is larger at one end than the other to fit the tapered arbor extension. Make sure any type of gathering pallet is correctly positioned. The long English pallet must rest on the pin at the end of the rack when the clock has struck and the train is locked—in fact, the train will not lock unless it is there. The small French pallet should be well clear of the rack at the same time—it needs a free run when the strike starts after the warning so that the train is running at its proper speed by the time the bell or gong is struck. The proper position of the motion wheels in relation to each other is essential if the strike is to work properly. You may have marked the wheels before stripping, but sometimes you can only discover the correct setting by experiment. Often, especially in French clocks, it is indicated by a punched dot between two wheel teeth, and a bevelled edge to the pinion leaf which must go between them.

In a quarter-striking carriage clock, examine the mechanism which obstructs or releases the hammer for striking the second bell. This is a lever which is raised slightly by the hour rack at each quarter (although that rack will not fall except on a 'grande sonnerie' clock) so that a pallet on the end of the hammer's arbor is free to drop into a notch on the lever and the hammer can strike. When the hours are struck, the rack is of course down and the lever will have fallen with it so that the notch no longer coincides with the pallet and this higher-pitched gong and hammer cannot strike (see Fig 38). This is a

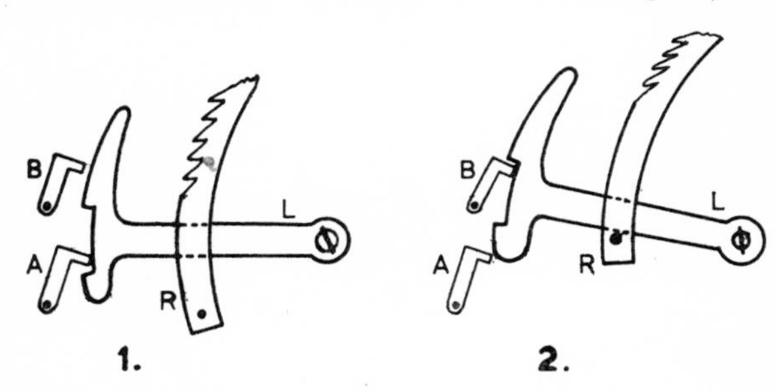

1. STRIKING THE HOUR. THE HOUR RACK 'R' HAS FALLEN, ALLOWING LEVER 'L' TO FALL. THE HIGH-NOTE GONG CANNOT BE STRUCK AS PALLET 'B', ATTACHED TO THE HAMMER ARBOR IS OBSTRUCTED BY THE TOP OF LEVER 'L'. LOW-NOTE HAMMER PALLET 'A' IS IN A SLOT IN THE LEVER AND IS FREE TO STRIKE.
2. STRIKING QUARTERS. THE HOUR RACK 'R' IS HELD RAISED BY A LEVER BENEATH IT DURING QUARTER STRIKING, AND LEVER 'L' ALSO IS RAISED. THE HIGH-NOTE HAMMER PALLET 'B' IS FREE TO FALL INTO THE SLOT ON LEVER 'L' AND SO STRIKE, AND THE OTHER PALLET 'A' ON THE LOW NOTE HAMMER, IS ALSO FREE.

Fig 38 Striking control in quarter-striking carriage clocks

difficult action to explain or illustrate, but observation and experiment will show you soon enough how it works. In this may appear the exception which proves the rule (though it does not affect the setting up); when the hours are struck on these clocks, the higher-noted hammer will in fact be 'up', since it cannot fall into the slot on the lever until the lever is raised by the pin on the end of the strike rack at quarter striking. But in fact, of course, it is not an exception, for the hammer is held up by the special lever, not by the pinwheel to which the rule applies.

In this type of clock ensure also that the lever, which normally prevents the hour rack from falling at the quarters, is moved aside at the hour so that the hour rack falls. This must happen when the quarter rack is on the highest segment of its snail or the clock will mix hours and quarters in a perplexing manner. A similar point has to be watched with ting-tang movements having only one rack in which a lever prevents the quarter hammers from falling during the hour striking. The motion wheel which operates this lever by pins must be so placed that the lever is fully raised just before the hour is due, and is free to fall, releasing all the hammers, before the first quarter is struck. There are often more than two hammers in these clocks but they are still classed with ting-tangs since there is no separate chiming, as opposed to striking train. Where, in more modern clocks, there is a chiming train working on a countwheel with the strike being on the rack principle, the position of this countwheel on its arbor is the critical factor. It is usually adjustable with a set-screw. A device such as a hump on the countwheel releases the hour train at the hour.

Setting up the mainspring

If you have a spring-driven fusee clock, the spring has to be set up. This is the time when you get rid of those coils of gut trailing round your feet, or fit the chain. Loosen the click of the barrel itself and move it away from the click wheel. Turn the barrel arbor slowly and guide the gut or chain so that it coils neatly and evenly round the barrel. Wind up all the gut onto the barrel. If there is a chain, wind most of it on and hook the remainder in position at the bottom of the fusee. At this stage you will have to hold the gut or chain taut so that it does not slide off the barrel. However the fusee is connected to the barrel, once the connection is taut grip the square of the barrel in a hand-vice or large key and wind up the spring between a half and one complete turn and, whilst you keep the tension on, slide the barrel click into place on the click wheel and tighten the fixing screw. The spring will now be set up so that its minimum power is not used and so that chain, gut or wire is held taut and will not slide about on the barrel.

Completing the assembly

Put the pallets into place or fit any platform escapement and adjust its depth to the contrate wheel teeth. You can now wind up the trains. As you wind the fusee going train, make sure that the fusee poke really does engage with the sprung stop-work detent when the gut or chain is nearly all on, and also ensure that the gut winds off the barrel evenly into the grooves of the fusee. When it has done so once, it will continue to do so thereafter.

Checking the striking

With the trains wound, you can run a full twelve-hour test on the striking system; this will be better done with the dial off, though that may not be possible in a French clock where the pallets swing in front of the dial and the scapewheel is inserted from the front. If the rack does not fall squarely onto each division of the snail—and if, in particular, the clock fails to strike twelve because the rack falls too soon onto the high step of the snail and one is struck—you will have to move the motion wheels in relation to each other until this point is corrected. Remember that in the countwheel system, whether English or French, the position of the countwheel is important. If the clock struck correctly originally, but is unreliable now you have cleaned it, moving the countwheel into a new position in relation to its arbor or driving pinion will probably be the cure.

Oiling

You must now do some oiling. Do not leave it until you have the dial on, because if you do you will not be able to do full justice to the many points on the front plate. Books have been written on the proper use of oil and chapters on its use in clock and watch movements. I am not going to follow the custom. There are only three things of paramount importance. The first is to use only clock or watch oil; the second is to apply it to all moving parts (except for the train teeth) where they touch any other part, and the third is to go easy with the quantities. You must have a small drop of oil—applied on the clean end of a needle, for example—on every pivot, every mounting stud, a minute drop on the crutch where it rubs the pendulum, rather more on the click-work, and a spot on the hammer tail where it contacts the pins. There are countless other places too, but they will be obvious to you without being listed. You must, on the other hand, keep oil away from a hairspring like the plague and apply only an extra-small drop to one or two of the scapewheel teeth. Again, while you must lubricate each pivot (a drop on the end will run into the hole), and whilst the cupped oil-sinks do retain oil, you do not want sinks filled to the brim, because the chances are that it will run down the plates, ruin the appearance, and be wasted. The train of wheels will in the long run be damaged by congealed oil and dirt and so should not be oiled. Over-oiling, particularly on a delicate escapement, can do great damage, but hardly more than the long-term damage caused by failing to oil; moderation and common sense must be your principal guides.

Hands and alarm

The final point of assembly will be fitting the bell or gong, if they attach to the movement, and putting on the dial and hands. If the minute hand is pinned on, devote a little time to finding a nice pin just the size of the boss, and file its end smooth so that there is not an ugly burr where the nippers have been. If the clock has an alarm, wind it and turn the set-square until the alarm goes off, then stop immediately. The time now shown on the main dial is the time to which the alarm hand must be set when you press it on. Ensure that the hour and minute hands are properly placed. Turn the arbor until the clock strikes the hour and then fit these hands exactly at the appropriate position for the hour struck. Occasionally a minute hand has to be moved on its collet so that striking really is right on the hour. Make sure that the hands do not cross each other when going round and also that the domed washer in front of a pinned minute hand is of a thickness to tension the spring against the cannon pinion below; a very stiff hand is liable to become bent and a loose one will have its own ideas on the time.

SETTING UP THE CLOCK

Putting in beat

The remaining major adjustment is setting the escapement 'in beat'. You can do it very roughly at this stage without the pendulum, or mount the movement on a level bracket and suspend the pendulum, but in any case a finishing adjustment will be needed when you put the movement back into its case. It is essential for the escapement to be in beat if it is to keep time and very nearly essential if it is to go at all. It is in beat when it produces a good solid 'tick-tock'—as near to a 'tock-tock' as can be arranged. More exactly, the pallets must act from the centre so that an equal impulse can be given to the pendulum or balance whichever pallet is releasing the tooth. They cannot operate like this if the balance is holding them crooked with its hairspring or if the pendulum is doing so through force of gravity.

A pendulum is usually set in beat by bending the crutch

slightly to one side or the other. 400-day clocks are adjusted by twisting the whole suspension (which is usually held by a screw or rivet friction-tight at the top) so that the angle of the forked piece clamped to the suspension spring is altered in relation to the crutch pin. The positioning of this forked piece is important vertically as well. When setting up, experiment with the effect on the escapement of moving the fork up and down the spring and then secure it in the best position. Too low will cause 'flutter'. Too high will cause the escapement to act jerkily and to stop. Two or three millimetres can be critical here. It is also essential that the fork be a free fit on the crutch pin. Some French clocks, as mentioned, have a friction-tight (or screwed) crutch which can be turned in relation to the pallets, but in others the crutch has to be bent (which is safer anyway). Most Vienna regulators have a device on the crutch whereby beat is set by turning thumbscrews; sometimes the crutch is in two parts whose angle to each other is adjusted, and sometimes the crutch pin which connects with the pendulum is itself moved by turning the screws.

Whatever the arrangement, the principle is always the same. An escapement is out of beat when, though the pallet arbor is at the correct height (ie the depth is all right), one pallet is too far from the scapewheel and the other is too near. If the clock does not 'tick' when the crutch and pendulum are over to the right, or ticks late, that pallet is too far off centre and the crutch will have to be bent towards the right (or, with a 400-day clock, the suspension and so the fork must be moved to point in that direction). When making the adjustment you must, of course, protect the pallets, and the suspension spring if it is in place. Therefore hold the top of the crutch firm, keep an eye on the spring (or remove if necessary), and bend gently with your other hand. It is best to assume that the clock-case is not itself crooked and will stand on a level surface, so you first set the clock in beat when it is standing level. If, then, the clock really must stand somewhere not quite horizontal, you will have to make a further slight adjustment to beat, for this is preferable to wedging the case on one side with bits of cardboard, pennies, and so forth.

A platform escapement will not of course be affected significantly by the clock's standing crooked, because the controller is positioned not by gravity but by the hairspring. To set the escapement you have to move the hairspring in relation to the balance for it will rarely be any use simply unpinning the hairspring from its stud and moving it in or out, because that will radically alter the time-keeping. You will have to take the balance and spring off the platform and unpin the end of the spring, removing it from the hole in the stud. Then—a delicate operation—loosen the spring's collet mounting on the balance staff by inserting the blade of a fine screwdriver in the slot, and turn the spring slightly on the staff without actually taking it off. Pin up the spring, replace the balance and cock, and observe whether or not the pallets are now central to the scapewheel. If not, repeat the process, making further adjustments. You may have to do it several times, but it is essential because a balance-wheel escapement which is out of beat really does find it very difficult to keep going at all since the impulses given through the lever to the balance are, at best, so small.

I do not want to labour the importance of a clock's being in beat. The fact is, however, that the subject is of particular importance to the ham. When the nature of his obsession is known, he receives frequent invitations to comment on the superb or deplorable condition of the clocks by which society is kept to order. A very large proportion of these clocks will be struggling pathetically, whatever their owners may think of them, not because of a radical sickness, but simply because they are out of beat. You will meet the lofty grandfather lollopping along as if one of its legs was wood, or be asked to admire the cherished French clock of the ancestral home which is obviously out of beat even from its appearance, because its round movement has been twisted in its case through repeated winding by untutored hands. Of course, sometimes nothing can be done to these handicapped tickers without giving offence to the proud possessor. You may spend a quarter of an hour on a tricky one (for it can be a very fine adjustment and awkward to get at inside the case), make it right, and receive only a remark that it has gone for years and there was nothing the matter with it. But the fact is that for any clock lover the heartbeat of a clock penetrates deep within him until it is put right. The confirmed ham may be so absorbed in the new clocks which come his way that he has no time to keep his existing collection properly cleaned, but one thing he will not permit is one of them to be out of beat.

Fitting the movement to the case

When the escapement is thus far in beat and the strike is reliable, the movement is ready to be mounted in its case. Very likely the case will have been attended to while some of the cleaning and other work was going on. If not, it must certainly be considered now for not only do the general state and finish of the case bulk large in the final appearance, but it is also important that the mounting of the movement be firm and true.

Long-case movements are almost invariably mounted on stout seat-boards sometimes, particularly with 'birdcage' movements, held in place by their own weight, but more often secured by stout hooks bent over the pillars and screwed up tightly underneath the board. The same method was employed in some of the older 'bracket' clocks, though during the eighteenth century they began to be mounted on stout brass brackets screwed into the sides of the case. With weight-driven clocks the rope, chain, gut, or wire (which I shall call 'gut' for convenience) has to pass through holes in the seat-board. Enlarge and move these if necessary so that the gut does not rub on the board and make sure you position the hooks with the same point in mind. Fix the loose ends of the gut, after passing them up again through the holes, on top of the seatboard. You can tie a large knot or bow, according to the size of the holes, but remember the strain on these fixtures. It is safer in the long run to make a peg out of stout steel wire and tie it firmly to the gut so that it lies flat across the hole. Then, with each train in turn, put the weights on temporarily or pull on the weight pulleys, and wind the clock up while guiding the gut evenly round the barrel. Test the length of a new gut against the case, from mounting-point to where the weight will hit the floor, and shorten it if necessary; the clock can only run at most until the weight contacts the floor, and if you have too much gut for the barrel it will tangle up and impair the transmission of power. So long as the gut is wound on evenly this time, it will stay that way on future windings. Wind the pulleys right up to the seatboard to keep the gut in place on the barrel (you should do this whenever you move a long-case clock) and screw the movement to the board by its hooks. Put the whole assembly on top of the 'cheeks' of the case and try the hood on; you want to avoid gaping gaps all round the edge of the dial but, on the other hand, you do not want the minute hand rubbing the glass or it will stop the clock. Make sure also that the dial squares up with the glass for a gap down one side is unsightly. Now reach into the depths of the case and hang the weights on their pulleys. If necessary pull on the lines while carefully releasing the clicks so as to make the pulleys accessible. The heavier of the two weights, remember, is for the striking train. If the case and cheeks are level, the crutch should start swinging to and fro at this stage. Finally mount the pendulum inserting it carefully through the crutch, if it is a closed-ended fork, and placing its spring between the suspension chops. The clock should still be in beat, subject to the case. But, if you have to wedge the seatboard up on one side to square the dial and hood, you will certainly need to take the hood off again and readjust the beat. Give the pulleys, and the crutch where it contacts the pendulum, a little oil, and the assembly is complete.

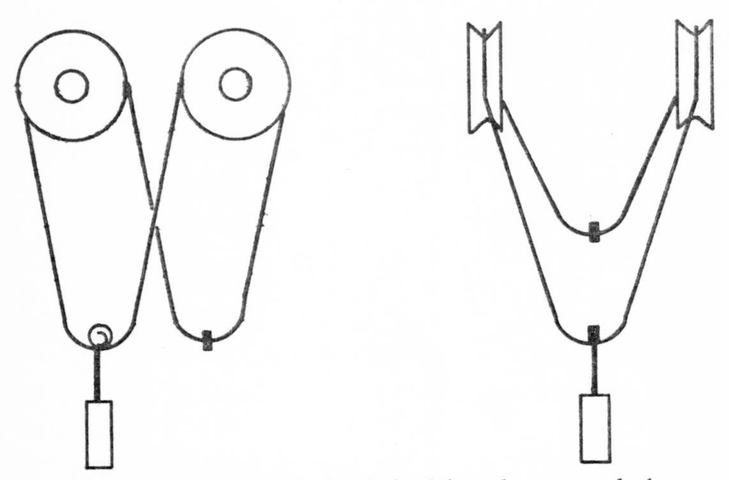

Fig 39 Position of chain in chain-driven long-case clocks

There is no essential difference when setting up a thirty-hour chain clock but you must take care to have the chain going over the driving pulleys in the right direction (see Fig 39), and make sure, before you close the joining link of chain or start trying to splice the rope, that you have not left out the counter-weight because this is essential if you are not to have earthquakes in the night as the chain slips and the weight falls.

You may now find that the pendulum swings so merrily that it collides with the side of the clock-case. It cannot happen with a properly adjusted escapement and original case, but in other circumstances it may. There are four solutions and you must try them and decide which, or which combination, works best. You can make the escapement less deep, fit a stiffer suspension spring, make the bob heavier, or use a lighter driving weight. If all else fails it is, of course, also possible to excavate into the thickness of the trunk's wood to provide more clearance for the bob, but this is a drastic and negative answer.

Positioning of the clock

The case must now be settled where it is going to stand, and the escapement set in beat for that particular location, which may not be exactly level. Ideally, there is no doubt that long-case clocks should be firmly screwed to the wall. But this is not always convenient in the modern home and it is not essential so long as the floors do not dip greatly away from the walls. If your clock is not screwed, it must be driven hard against the wall by pushing small wedges under the front edge of the base. If there is a thick skirting-board to be negotiated, a batten of the same thickness as the existing board fitted to the back of the case well above skirting level will get round the difficulty. Some of the older clocks had bases, and were apparently made so, with cut-outs at the sides for this reason, but it is unnecessary to cut up a good old case without them in the same way.

It is just as essential to mount a Vienna regulator or similar clock firmly, screwing it to the wall with screws as large as the mounts will permit. I had one once on a wall which I did not want to mark permanently, and tried to suspend it from a chain fastened between two door-jambs. The arrangement was not a success. The whole clock came down as it was striking and the case had to be rebuilt, though fortunately only the pendulum of the clock itself was damaged. The best of these clocks often have thumbscrews mounted in brackets at the bottom corners of the case to adjust for irregularities in the wall. The dead-beat escapement operates with a very small arc of the pendulum and has to be properly aligned and set well in beat.

The purpose of fixing up a long-case clock firmly is not

simply so that it will not be knocked over, although that is a point to consider. The currents of air set up by these large pendulum bobs, and of course the effect of counter-motion, can cause the weights to swing also, and then the whole clock can become so unstable that it will keep bad time, stop, or worse. Persistent trouble from swinging weights which hit the bob can often be cured by flattening (ie streamlining) the bob or, in the worst cases, by arranging a screen across the case half way up between pendulum and weights.

Repairing cases

The full restoration of an old long-case clock case is work for a cabinet maker or the few amateur clock makers who possess such skills. But many of these cases will come to you in an appalling condition and their basic structure can be fairly easily restored; such work usually depends on no more than the glueing and pinning of thick timbers at right-angles. Signs of woodworm must be treated with proprietary fluid or spray. Woodworm is a menace to the clock, all other furniture and to the house itself. Pay especial attention to the condition of the cheeks and backboard. A backboard which has warped out of the true must be replaced or, if it is not too bad and the pendulum does not hang from it, it may be straightened by screwing substantial battens across it. The same applies to a backboard split down the middle. Replacements should be in seasoned wood, preferably hardwood. It is worth taking trouble here, especially if the house is centrally heated. Cheeks which are seriously out of true must be straightened with sandpaper or a plane; a movement which rocks is no good to anyone.

The cases of 'birdcage' movements are simpler still. They are not usually elaborately finished and their construction is rugged rather than skilful. The supply of movements seems to have outstripped that of cases—probably because good cases are requisitioned for eight-day movements—but it is not a difficult matter to make up a case, either of deal or hardwood, choosing proportions which appeal to you and finishing with mouldings roughly appropriate to the date of movement and dial. A good 'antique' finish can be given by repeated applications of beeswax dissolved in a small quantity of real turpentine to which stain of the desired colour has been added. Three parts of turpentine to one of beeswax is a good mix. Melt the beeswax in a double saucepan or in a tin placed in a saucepan of water. An ebonised finish can be given by three good coats of blackboard paint followed by an application of the same polish. For convenience, when the required colour and surface have been achieved, the whole can be sealed with polyurethane varnish. Taste may, however, require that this final coat be sanded down with very fine paper and steelwool because a high gloss may not be 'right'. Alternatively, matt varnish is obtainable. Such a case will protect the movement and can be made of a shape and style you really like or, using a photograph as a guide, in scaled-up imitation of a case which attracts you. It will serve you well until you come across a more appropriate and contemporary housing. Of course these movements are robust and often of interestingly crude appearance, so it is possible to mount them caseless on stout brackets screwed to the wall where they can be seen, but they then need frequent cleaning.

Finally, it has to be said that, with all precautions taken, the guts, ropes, and wire of weight-driven clocks do break, and chains sometimes run foul of their pulleys. The load which hits the floorboards on these occasions is not to be underestimated. Few such cases have good, if any, timber at the base, and it is only sensible to put a thick cushion or pad of foam rubber into the bottom in case you have a fall. Apart from the aspects of craftsmanship and time-keeping, this is a reason to make sure of the suspension. A seconds pendulum falling three feet will go a good inch into many floorboards and the regulating threads will keep it standing there hard and fast. Drilling one out makes a mess of a strip floor as well as bending the pendulum.

The cases of spring-driven clocks are not subject to these particular dangers, but they also often arrive in abominable condition and these cases, with their finer movements, need to be a good deal more secure against invading foreign bodies. They also tend to be highly finished and are subjected to a cumulative strain arising from winding. It is necessary for brackets and fittings to be well screwed in with, if necessary, new screws and plugged or freshly tapped holes. Cracked glass panels need to be replaced and loose panels wedged with fine fillets of wood or with a little putty, not only to keep the dirt out, but also so that they do not rattle or squeak when the clock strikes. Gilt fittings can be washed in a soap solution, brushed clean and sent away to be regilded if they are valuable. If not they may be touched up with gold paint and wax or merely lacquered. Holes and dents in the woodwork need to be filled, stained and polished and veneers reglued, after being exposed with a razorblade, wherever possible. Behind the more elaborate frets, whether of metal or of wood, there is frequently coloured silk. It will usually be found that the original silk is not only dirty but also rotten. Replacing this with new material (which may be synthetic) will make a great deal of difference to the appearance.

Adjustment of bells and gongs

Many a clock, even one professionally restored, is not allowed to give of its best when it comes to striking. It is almost as if it is thought that the bell or gong clock apparatus is a bonus added to the clock for the benefit of illiterates who cannot read the dial. This is not so. The striking of a clock is an essential part of it, even though it is merely set off by the time-keeper and entirely dependent on it. Hammers which strike bells must strike them soundly, not from a crooked angle, and strike them as near to the edge as possible, whether they strike on the inside or the outside. The bells themselves must be screwed on really firmly and set the correct distance from the hammers when they are down. There are few sounds more horrible to the clock-lover than that of a hammer jangling on a gong or a bell which is too near to it. A clock hammer really falls not onto the bell but onto a stop, which may be a stout pin or a piece of metal or merely one of the movement's pillars. The sudden arrest of the shaft of the hammer causes it to bend momentarily and thus the hammer head itself strikes the bell and then springs smartly away from it. Alternatively, a rigid hammer falls onto a stiff spring. You can and should adjust the position of the bell to give a suitable volume, but do not put it so far away from the hammer that the tone is a mere apology or so close that the hammer hits it more than once for each stroke. You can and should also adjust the speed of strike, either by choosing a suitable weight, by altering the depth of the fly pinion, or by slackening or increasing the hammer spring pressure. On an old English clock the spring is usually a strip of steel. On a French clock it will be a thin steel wire screwed into the front plate; make sure that this is behind the hammer lug which usually has a little slot filed in it to accommodate the spring and make the same check for the spring on the locking piece which will give trouble sooner or later if the spring is not acting on it correctly.

The noise of a gong can be even nastier than that of a bell if it is not set up properly. The sort of gong which is screwed into a block on the end of a post fitted to the case must be as firm as it can be. If the nut which tightens the post is small, replace it with a larger one and add a sizable washer for good measure. Make sure that the pendulum cannot contact the gong. It is common, but quite unnecessary, for the pendulum to rub the gong slightly as it vibrates when striking. You will know from when you dismantled the clock that wire gongs themselves make rather a thin noise; the tone and body of the strike depend on utilising the resonance of the case, and this cannot be done unless the gong is really well mounted on it. The hammers usually have a soft spot which hits the gong—this is a block of leather pushed or glued into the hammer head. If the spot is hard from thousands of impacts or has a layer of grease on it, scrape it clean with emery and if necessary pierce it all over with a sharp needle. It must not be fluffy, but the gong will not give of its best if it is being hit by a really hard lump. Many gongs have no very impressive sound at the best of times, but it is worth doing what you can for them.

Setting up carriage clocks

All these adjustments are particularly important to carriage clocks, where the wire gongs are in a brass block or blocks screwed to the back plate. The hammers are stopped by wire pins and the best possible distance of the head from the gong—which depends on the flexibility and length of the hammer shaft—must be found by experiment. Make sure that these tiny gongs are hit close to their mounting point and square in the middle of the wire. A gong hit too far along will merely rattle. These wire gongs are almost always blued and reblueing them is not easy, because the tone will be spoilt if you uncoil them when cleaning them and excessive heat when blueing them will produce the same result. Do not, however, be tempted to give a wire gong a coat of blue enamel or its note will be reduced to an unmusical thud and, though you may wipe it with an oily rag, do not flood it with oil.

The tone of a striking carriage clock is considerably dependent upon the surface on which the clock is standing. There is very little resonance in so small a metal case alone. But this does not mean that looseness in any part of the case can be tolerated. In particular, a loose glass panel, which is common enough on the best of carriage clocks, will ruin the tone of the striking. Do not use glass cement to rectify this, for it will mess the glass and make the case impossible to clean properly in future. Loose glasses should be fixed wth slips of cards or cork pushed down between the pillars and the panel edges.

Another small point to watch with carriage clocks is the fit of the door and carrying handle. The early solid brass doors are usually a good fit but the more common doors consisting of a glass panel in a brass surround, often fit very badly and are not even square at the corners. There is not much that can be done, unless you have the facilities for making a new side to the door from square-sectioned brass, but it is worth bending the pins which form the hinge so that the best compromise fit is obtained and there is not a large gap on one side. You will find that there is a certain amount of give in the brass frames of these cases and, if you loosen the pillar screws, you may be able to improve the fit by moving the pillars very slightly. The top panel of glass, above the escapement, must be firm, but do not screw it so tightly that the glass becomes chipped or cracked by its securing clamps. You have to make another compromise here since the tightness of the glass is also controlled partly by the spring of the handle. Matters should be adjusted so that the handle is sprung between the turned pillars as the top-panel fixing screws are tightened. A handle does not, in my view, look well flat on top of the case where, if it is loose, it may rattle. The spring should be such that the handle will stay up of its own accord.

Setting up French clocks

The marble case of a cheaply bought French clock commonly reaches you with a very unattractive appearance. Do not use an abrasive on it, but give it a thorough soapy wash and finish it with a stiff polish of beeswax and turpentine. Hard manual labour is needed to bring up the shine but the result, if you have not tried this treatment before, will surprise you. Alternatively, rub hard with metal polish, clean off with white spirit and finish with clear lacquer. The engraved lines are sometimes left clean (in which case, scrape the polish from them) but are more often given a gold finish, and gold paint or wax will bring out the markings well. Cracks can be filled with plaster of Paris or a resin filler, stained if necessary. Once it has been sanded really smooth and level and well polished with beeswax the repair will hardly be visible.

In a well-designed French clock, however elaborate the case-work, the dial dominates the impression and an important element here is the brass bezel. It will need a very thorough clean, possibly with the wire brush and certainly with metal polish. Then it must have a coat of lacquer or it will quickly return to its undistinguished brown.

Remember that the mounting of French clocks is almost always a form of clamp, the case being gripped between the front and back bezels. Make sure you set up the clock level and with the dial straight, and then tighten the fixture well, for the design makes these movements especially susceptible to the strain of winding, and they need to be exactly in beat.

9

The end of it all

You have now acquired, and possibly cleaned and repaired, a clock of some distinction or interest. Perhaps it is your first, looked at from the special viewpoint of hamhood, or perhaps it is one of many. Its purpose will vary—it may be ornamental, it may gratify your ego, it may astound your neighbours, it may be held with the intention, if not the actual prospect, of ultimately making a small fortune. I have a good electric master-clock which drives a couple of slaves in different rooms, at whose dials I never look, but whose real purpose is to startle unwanted visitors with the loud 'clunk' it makes every thirty seconds.

Whatever the clock's real purpose, its ostensible purpose is to tell the time, and for a while that will seem very important. For a few days, perhaps weeks, you will study the face at regular intervals and need a keen eye to detect minute variations from your wristwatch which may itself be no remarkable time-keeper. Then you will move into higher regions of precision and set the clock by the golden voice of the telephone or the signals of radio and television. You will adjust the pendulum's rating-screw or the index of the balance-wheel time after time, using perhaps a schedule by the clock on which to record the last modification made. In the early stages, you will no doubt play with the escapement or alter the driving weights. But, in due course, you will conclude, with countless others before you, that the dial of your clock was not drawn in a very sophisticated manner in the first place, or that to tamper with so ancient a mechanism would be to violate some inherent virtue inseparable from antiquity. Alternatively, you may decide that the climate is unsuited to uncompensated controllers, or that central heating makes a nonsense of mechanical exactitude.

So, bit by bit, though you will be annoyed by the vast majority who foolishly expect a clock to tell the right time rather than respecting the more considerable miracle that it actually keeps going, you will either lower your standards or come round to the seductive notion that all clocks are, in reality, set to time by other clocks and devil take the hindmost. That austere codger in the corner is correct in its own small world and that is what we expect in a democratic society; a clock has the right to hold a divergent opinion so long as it does not seek to impose its view on the massive mediocrity which is of the contrary persuasion.

Nonetheless, the austere codger, though it will do its utmost on its great bell to draw attention to itself, will slowly sink into qualified oblivion even if its performance is flawless. Its proportions, once cherished every time you passed the door and looked in, will gradually become part of the general décor. The wife will, before long, sleep through the clamour at night and the personage, like others in the family, will eventually receive only passing notice unless it is taken sick. Then, perhaps, you will doctor it, give it a thorough cleansing all over again and heal its lesions. But only in exceptional instances will you do so with quite the initial enthusiasm.

Other than in a financial extremity, this is the first time when you are likely, perhaps after several such operations, to harden the heart and to let the guilty thought of a change cross your mind. For if the clock which performs perfectly month after month does not satisfy every desire, neither does the clock which, after all the ministrations, is still unreliable.

To salve your conscience possibly, to get a higher price more probably, you will polish the now-tarnished movement till it is again resplendent, make such adjustments as are feasible and prepare a story to persuade others of the clock's substantial merits. It is then that you may realise painfully what you always secretly suspected, that the original price was slightly over the top in the first place for reasons which will not now cut much ice with a prospective purchaser. But eventually you will, with luck, at least break even. More to the point, you will know that, before very long, the proceeds will be the means of again bringing you a fresh clock with all its own subtle snares and novel delights.

Index